Random Finite-Valued Dynamical Systems: Additive Markov Chain Approach

The Kharkov Series in Physics and Mathematics

Introduction to the series

This new international book series entitled *The Kharkov Series in Physics and Mathematics* provides a unique medium for the publication of high quality original monographs, multiauthor volumes and reference texts in priority research areas of physics and mathematics for graduate, postgraduate and research/professional readership. The Institute for Low Temperature Physics and Engineering of the National Academy of Sciences of Ukraine, Kharkov has had a long and outstanding tradition in research in physics and mathematics, and is particularly associated with the classical studies and publications of L.D. Landau, L.V. Shubnikov, M. Ostrogradskii, A. Lyapunov and V. Steklov and more recently with the achievements of I.M. Lifshitz, A.I. Akhiezer, V.A. Marchenko, A.V. Pogorelov and V. Drinfeld. Continuing this tradition, this book series will publish and highlight research areas including: condensed matter (theory and experiment), superconductivity, low temperature magnetism and interaction of radiation with matter, cryocrystals, point-contact, optical and magnetic resonance spectroscopies, quantum phase transition, geometry and topology, complex analysis, mathematical modeling, statistical methods in mathematical physics.

Random Finite-Valued Dynamical Systems: Additive Markov Chain Approach

O.V. Usatenko, S.S. Apostolov,
Z.A. Mayzelis, and S.S. Melnik

Institute for Radiophysics and Electronics
National Academy of Sciences of Ukraine,
Kharkov Ukraine

Cambridge Scientific Publishers

Cover design: Clare Turner

British Library Cataloguing in Publication Data
A catalogue record for this book has been requested

Library of Congress Cataloguing in Publication Data
A catalogue record has been requested

ISBN 978-1-904868-74-3 Hardback

Cambridge Scientific Publishers Ltd
P.O.Box 806
Cottenham, Cambridge CB24 8RT
UK
www.cambridgescientificpublishers.com

Printed and bound in Great Britain by
CPI Antony Rowe, Chippenham and Eastbourne

Contents

1. Introduction

A dynamical system consists of a state determined by a set of points or a vector belonging to an appropriate state space (or an abstract phase space), whose coordinates describe dynamical state at any instant. Definition of dynamical systems also includes the rules describing the way that the state of system undergoes a change through time. The evolution rule of the dynamical system is a fixed recipe that prescribes the future states following from the current state or from a series of states from the past.

Dynamical systems are "deterministic" if there is a unique consequent to every state, and "stochastic" or "random" if there is more than one consequent chosen from some probability distribution or determined by solution of a stochastic differential equation. A dynamical system can have either discrete or continuous time. A dynamical system is called discrete if time is measured in discrete steps; these are modelled as recursive relations. If time is measured continuously, the resulting continuous dynamical systems are expressed as ordinary differential equations. The continuous case is defined by a "flow" and the discrete case is defined by a map. The goal of the present review is to study the *discrete stochastic* dynamical systems. Special attention will be paid to systems with long-range spatial and/or temporal correlations (LRCS). It is one of the topics of intensive research in modern physics, as well as the theory of dynamical systems and the theory of probability. The well-known examples also include cellular automata and Boolean networks, which have found broad applications in engineering, computer science, and, more recently, computational biology. These systems are wide-spread in physics [1]. They are gravitational forces [2], Coulomb forces in globally charged systems [3], wave-particle interactions, magnets with dipolar interactions [4], etc. These kind of systems are very often characterized by non-Gibbsian distributions, e.g., Zipf distribution in linguistic [6, 5], the distribution of nucleotides in DNA sequences and computer codes [8, 7, 9], the distributions in financial markets [11, 10, 12], other distributions in sociology, physiology, seismology, and many other sciences [13, 14]. The LRC systems are usually characterized by a complex structure and can contain a number of hierarchic objects as their subsystems.

At the present time, there is no generally accepted theoretical model that adequately describes the dynamical and statistical properties of the random LRC systems. Attempts to describe the behavior of the LRCS in the frame-work of the Tsallis non-extensive thermodynamics [15, 16] were made. However, the non-extensive thermodynamics are not well-grounded and require the construction of the additional models which could clarify the properties of the LRC systems.

One of the efficient methods to investigate the correlated systems is based on a decomposition of the space of states into a finite number of parts labelled by definite symbols. This procedure referred to as coarse graining is accompanied by the loss of short-range memory between states of system but it does not affect and does not damage its robust invariant statistical properties on large scales. This naïve view on robustness will be discussed in more detail in Subsection 4.6. Initially the procedure of coarse graining was proposed by

P. and T. Ehrenfests [17], who put forward the idea of creating the irreversible dynamics using coarse graining. This method can be applied to examine both discrete and continuous systems. As a result we obtain a symbolic LRCS. Yet it is much easier to investigate the numeric LRCS, thus the next step towards simplification of the problem is to map the symbols into numbers.

The most frequently used method for decomposing the set of symbols or space of states or phase space is based on the introduction of its two parts. In other words, it consists in mapping the two parts of states onto two symbols, or numbers, say 0 and 1.

One of the ways to get a correct insight into the nature of correlations consists in an ability of constructing a mathematical object (for example, a correlated sequence of symbols) possessing the same statistical properties as the initial system.

In many areas of physics such as engineering and signal processing there arises nearly the same problem: generation of colored noise, i.e. a correlated random process. Since the pair correlations usually make the principal contribution to the observable quantities, the problem of a random sequence generation with the prescribed *pair correlation function* is of particular interest. It has been known for a long time that continuous colored noise with exponential correlations is generated by the linear Ornstein-Uhlenbeck process (see the modern review in Reference [18]). A more general method valid for generation of continuous random sequences with an *arbitrary* correlation function, is based on the convolution of white noise with the *modulation function* defined by the pair correlation function. Originally, the convolution method was proposed by Rice [19]. Applications of some versions of the method with specific correlations, including the long-range non-exponential correlations, can be found in References [20, 24, 21, 25, 23, 26, 22]. Among other methods of colored noise generation we can emphasize Mandelbrot's diffusion [27], the Voss procedure of consequent random additions [28], the expansion-modification Li method [29], and the correlated Levy walks [30]. We believe that, among the above-mentioned methods, the use of the Markov chains is of great importance.

One-dimensional LRC systems are either (discrete) chains or (continuous) processes. The first of them are the discrete sequences of elements while the latter are the random variables depending on the continuous "time"-like argument. The processes are of great interest in physics, but the practically used method consists in discretizing the process. As a result of such an approximation we attain a chain, and thus the investigation into the problem of the random chains is an important one.

1.1 VARIETY OF MARKOV CHAINS

To define a random chain (or any other random object) we should present the probabilities of occurring different states of the system. For the chain, this means the probabilities that the different symbols take on some fixed values are given. Furthermore, the probabilities of all the pairs of symbols at different distances from each other are to be introduced, the same is applied to all the triples of symbols, and so on. Note that the probabilities for the pairs of the symbols at the same distance between each other but located in the different regions of the chains are not necessarily equal (in this case the chain is not homogenous). Such a straightforward method of the random chain definition is rather cumbersome because in the general case one has to prescribe a great number of probabilities for the different many-symbol words. Yet another approach to the random chains is based on using the conditional probabilities. To consider an object as a Markov chain is an example of such an approach.

The *N-step Markov* chain is a sequence of random variables a_i, $i \in \mathbb{Z} = \{\dots, -2, -1, 0, 1, 2, \dots\}$, possessing the following property: the probability of symbol a_i to have a certain value under the condition that the values of all **previous** symbols are fixed depends on the values of N previous symbols only,

$$P(a_i = a | \dots, a_{i-2}, a_{i-1}) = P(a_i = a | a_{i-N}, \dots, a_{i-2}, a_{i-1}). \qquad (1.1)$$

We refer to N as *the memory length*. In what follows we will use notation $T_{N,i}$ for the N-word preceding symbol a_i, i.e. the set of symbols $a_{i-N}, \dots, a_{i-1}$. The Markov chain is a *homogeneous* sequence, because the conditional probability Eq. (1.1) does not depend explicitly on i, i.e., is independent of the position of symbols $a_{i-N}, \dots, a_{i-1}, a_i$ in the chain. It depends on the values of $a_{i-N}, \dots, a_{i-1}, a_i$ and their positional relationship only. If the conditional probability does not depend on the order of symbols in the conditional probability function and is dependent on the number of unities in it we referred to it as a *permutative* Markov chain.

An important property of a correlated sequence is ergodicity. We suppose that the conditional probability $P(a_i \mid T_{N,i})$ differs from zero and unity for any word $T_{N,i}$ that provides the metrical transitivity of the Markov chain (see Appendix A). According the Markov theorem, this property, in its turn, leads to the ergodicity of the symbolic system under consideration. The infinite chain with this property contains all possible combinations of symbols, i.e. the state space is compact. Starting from some combination of symbols and building the chain according to the conditional probabilities Eq. (1.1) we have a non-zero probability to get any other combination after constructing some symbols. That, in particular, means that the chain is homogeneous provided that conditional probability function does not depend on index i explicitly. In what follows we suppose these conditions to be fulfilled.

An important class of the random sequences is the *binary* chains. If each symbol a_i of the chain can take on only two values, s_0 and s_1, then we refer to this chain as a binary one. It is convenient to change a value of a_i to 0 and 1 using a linear transformation,

$$a_i := \frac{a_i - s_0}{s_1 - s_0}.$$

Still more particular but just a highly important class is the *additive* binary Markov chains. The conditional probability of these chains is described by the following formula (this expression is substantiated in Subsection 2.2,

$$P(a_i = 1 | a_{i-N}, \dots, a_{i-2}, a_{i-1}) = \bar{a} + \sum_{r=1}^{N} F(r)(a_{i-r} - \bar{a}). \qquad (1.2)$$

Here function $F(r)$, $r = 1, \dots, N$, is referred to as the *memory function* (MF) (see Reference [31]) and $\bar{a}$ is the average number of unities in the sequence. The additivity property of the Markov chain implies that the influence of previous symbols on a generated one is independent. The concept of additivity, originally introduced in paper [32], was later generalized to the class of binary non-stationary Markov chains [33].

The latter were defined in [33] by the following conditional probability function,

$$P(a_{L+1} = 1|a_1, \dots, a_{L-1}, a_L) = \frac{1}{2}\left(1 - \mu\frac{L-2k}{L+L_0}\right). \tag{1.3}$$

Here k is the number of unities among the first L symbols, and parameter $-1 < \mu < 1$ describes the correlations strength in the chain, and length L_0 is the characteristic length at which the correlations begin to play the crucial role. These chains are non-stationary due to the dependence of the conditional probability (1.3) on the position L of the symbol generated and thus are non-ergodic. That means, in particular, that averaging over the ensemble of such chains is not equivalent to averaging over the length of the chain. The chains determined by the conditional probability (1.3) are shown to undergo a dynamical phase transition from natural diffusion into a superdiffusion phase at the critical value of parameter $\mu_c = 1/2$. The natural diffusion implies the linear growth of variance $D(L)$ with the string length L while the superdiffusion is characterized by variance $D(L) \sim L^{\alpha}$ with parameter $\alpha > 1$. This model is useful for the analysis of different non-stationary processes and non-ergodic systems. However, there exists a vast class of stationary systems (for example, written texts or DNA sequences) that should be modelled by the stationary Markov chains.

Another generalization is based on consideration of the Markov sequence idea with a multi-valued alphabet [34, 35]. The authors investigate such chains in the mean-field approximation, introducing the following conditional probability of finding zero at ith site,

$$P(a_i = 0|a_j, j \neq i) = \frac{1}{N}\sum_{j=1}^{L-1} n_j g_j. \tag{1.4}$$

Here N is the length of the L-valued chain, n_j is the number of the jth symbol ($\sum_j n_j = N$), and g_j is the *a priori* probability of choosing this symbol. This model is demonstrated to be exactly solvable. The procedure of coarse graining of such sequences is thoroughly investigated. It was proven that the coarse graining affects essentially the correlations in the chain. Specifically, the sequence showing diffusive behavior can reveal super-diffusive behavior after the coarse graining. Further, we shall discuss the problem of coarse graining of written texts and DNA sequences.

The function $P(a_i = 1 \mid T_{N,i})$ contains complete information about correlation properties of the Markov chain. In general, the correlation function and other moments are employed as the input characteristics to describe the correlated random systems. Yet, the correlation function takes account of not only the direct interconnection of the elements a_i and a_{i+r}, but also their indirect interaction via other elements. Our approach operates with the "origin" characteristics of the system, specifically with the memory function.

1.2 STATISTICAL CHARACTERISTICS OF RANDOM SEQUENCES

Along with the conditional probability, which indicates the direct association between the symbols, there are the high-order correlation functions. In the theory of spatially disordered systems the role of the pair correlation function (for random potentials) is due to the fact that it determines the scattering cross section in the Born approximation (for weak potentials). As a result, many linear transport characteristics (conductance, transmission and reflection coefficients, localization length, etc.) are expressed via the pair correlation function. [36, 37]

A correlation function of the sth order is defined as follows:

$$K_s(r_1, r_2, \dots, r_{s-1}) = \overline{(a_0 - \bar{a})(a_{r_1} - \bar{a})(a_{r_1+r_2} - \bar{a}) \dots (a_{r_1+\dots+r_{s-1}} - \bar{a})}. \tag{1.5}$$

Here notation $\overline{\dots}$ is the statistical average over the ensemble of chains.

Formally, function K_s has to depend on s arguments (the indexes of the symbols), but the considered Markov chain is a stationary one. Function K_s then depends on $s-1$ arguments (difference between the indexes of the neighboring symbols $r_1, r_2, \dots, r_{s-1}$). The correlation functions point to the implicit interaction of the symbols. If we specify values of the correlation functions of all orders, we determine completely the random sequence. So, it is an important problem to find the high-order correlation functions. The binary correlation functions $K(r)$, that corresponds to the pair interaction between the elements of the chain are especially significant. It is defined as follows:

$$K(r) = K_2(r) = \overline{(a_0 - \bar{a})(a_r - \bar{a})} = \overline{a_0 a_r} - \bar{a}^2. \tag{1.6}$$

By definition, the correlation function is even, $K(-r) = K(r)$, and the variance of the random variable a_i for the binary chain is $K(0) = \bar{a}(1 - \bar{a})$.

In order to study the statistical properties of the Markov chain, we consider the distribution $W_L(k)$ of the words of definite length L by the number k of unities in them,

$$k_i(L) = \sum_{l=1}^{L} a_{i+l}, \tag{1.7}$$

and the variance of k,

$$D(L) = \overline{k^2} - \bar{k}^2. \tag{1.8}$$

If the chain is non-correlated and contains the equal number of zeros and unities we arrive at the known result for Brownian diffusion,

$$D(L) = L/4. \tag{1.9}$$

As it follows from the definitions Eqs. (1.6) and (1.8) the variance is connected with the binary correlation function by the equation

$$K(r) = \frac{1}{2}(D(r-1) - 2D(r) + D(r+1)), \tag{1.10}$$

or

$$K(r) = \frac{1}{2}\frac{\mathrm{d}^2 D(r)}{\mathrm{d}r^2} \tag{1.11}$$

in the continuous limit.

1.3 MODELS INVOLVING MARKOV CHAINS

Here there should be mention of the Hidden Markov Models that allow us to construct or/and study a correlated sequence, which is not a Markov chain, by means of the Markov chains. The Hidden Markov models (HMMs) were originally introduced in the statistics literature as far back as 1957. Subsequently, their use was partially successful in a variety of applications in the engineering world, starting in the late 1970s. These applications include speech, handwriting, gesture recognition [38, 39] and source coding. In the second half of the 1980s, HMMs came to be applied to the analysis of biological sequences, in particular DNA. Since then, they have become ubiquitous in the field of bioinformatics (see, e.g., [38, 40, 41, 42, 43]). In recent years, HMMs have also been used to resolve some problems in computational biology, such an identifying the genes of an organism from its DNA [46, 44, 45] and classifying proteins into a small number of families [47]. The bibliographies of [48, 49] contain many references in this area.

A HMM is a statistical model where the system being modelled is assumed to be a Markov process with unknown parameters, and the challenge is to determine the hidden parameters from the observable parameters. The extracted model parameters can then be used to perform further analysis, for example, in pattern recognition applications.

In a regular Markov model, the state is directly visible to the observer, and, therefore, the state transition probabilities are the only parameters. In a hidden Markov model, there is the sequence of the states $\{x_n\}$, which is not directly visible, but the sequence of the symbols $\{a_n\}$ influenced by the states $\{x_n\}$ are visible. Each state x_n has a probability distribution over a set of the possible output symbols ⁀(for instance, ⁀$= \{0, 1\}$). Therefore the sequence of symbols $\{a_n\}$ generated by an HMM provides some information of the sequence of states.

As a simple example, the DNA sequences can be considered. It can be presumed that there is the prototype x_n of an existent DNA sequence a_n. Then there is the probability function $p(a_n|x_n)$ which codes certain x_n by one of the four DNA symbols. If the prototype $\{x_n\}$ is the Markov chain, this model is the HMM.

There are three canonical problems associated with HMMs.

1. Given the parameters of the model, compute the probability of a particular output sequence. This problem is solved by the forward algorithm [50, 51, 52].
2. Given the parameters of the model, find the most likely sequence of hidden states that could have generated a given output sequence. This problem is solved by the Viterbi algorithm [50, 53, 54].
3. Given an output sequence or a set of such sequences, find the most likely set of state transition and output probabilities. In other words, train the parameters of the HMM given a dataset of sequences. This problem is solved by the Baum-Welch algorithm [55].

The most natural and important instance in which we have to deal with probabilities is the measurement in a quantum system. Thus, the reformulation of the classical Markov chain in the quantum mechanical terms is the natural object of investigations. In recent years a problem of the so-called *quantum Markov chains* aroused considerable interest.

First of all it should be noted that there is no widely accepted definition of this object. The simplest of them is as follows. An infinite quantum chain is defined by a 1-D lattice, which is $\mathbb{N}$ (or $\mathbb{Z}$) and each site of the lattice contains a spin degree of freedom. So, the quantum Markov chain is analogous to the classical Markov chain except that the probability measure is replaced by a complex amplitude measure and transitional probability matrix is replaced by a transition amplitude matrix. Other definitions of a quantum Markov chain presented in Reference [56] is discussed at the end of Section 5.

Dichotomic symbols in a Markov chain can be thought of as the sequence of states of some particle, which participates in a correlated Brownian motion. Every element of the sequence corresponds to the instant change of the particle's coordinate. For example, we can associate symbol "1" with an increase of the particle's coordinate by 1 (a jump) and symbol "0" with the invariability of its coordinate (absence of the jump). Thus, every L-word (the sub-sequence of symbols of the length L in the sequence) can be regarded as one of the realizations of the ensemble of correlated Brownian trajectories over the "time" interval L. The positive values of the MF result in persistent diffusion where previous displacements of the Brownian particle in some direction provoke its consequent displacement in the same direction. The negative values of the MF correspond to the antipersistent diffusion where the changes in the direction of motion are more probable. In terms of the Ising long-range particles interaction model, which could be naturally associated with the Markov chains, the positive values of the MF correspond to the attraction of particles whereas negative ones conform to the repulsion.

1.4 STRUCTURE OF REVIEW

Now we give the brief content of the review. The second Section contains the development of the general approach to the Markov chains. We start from reviewing a particular case of an additive Markov chain with the step-wise MF. This case is of great importance because the explicit expressions for the distribution function $W_L(k)$ were obtained. After the detailed analysis of these chains we proceed to the study of a more general case of the additive but not only step-wise MF. We establish a very important relation between the MF and the binary correlation function of the chain. Using this relation we further express the correlation function via the MF. The studying an arbitrary Markov chain (not solely additive one) is completed by the next theoretical part. In different applications one encounters not only the objects that are characterized by the probability of a definite symbol on condition of the given values of the previous ones. There are systems where it is easy to find a probability on condition that the values of all surrounding symbols are fixed. Such chains referred to as two-sided sequences, are discussed in the next Subsection. They are proved to be Markov chain and their "one-sided" conditional probability is found. This equivalence is then used to study the isotropy properties of the Markov chain. We review the derivation of a set of equations that determine the correlation functions of different orders for the Markov chain with a given conditional probability function.

The third Section deals with methods of constructing the correlated sequences. First, we expound the Rice method of generating a correlated random series from an uncorrelated one. Further we discuss the method based on constructing the Markov chain with a prescribed binary correlation function. In the case of weak correlations it is shown that this method can be generalized to constructing the chains with prescribed correlation functions of the second and third orders. Then we suggest a method for generation of random binary sequences with prescribed correlation properties, method of filtering probability functions, Ref. [57]. Finally we analyze the signum-generation method. It is based on one of modifications of the convolution (Rice's) method of constructing continuous random processes. Apart from the theoretical interest, the methods considered can be used in various applications, e.g. the design of one-dimensional devises with the selective transport properties.

The last Section contains various applications of the developed theoretical tool. We start from the physical example of the Ising spin chain with long-range interaction. We build a non-extensive thermodynamics for such a system where the standard approach is not valid. The next system under consideration is a set of superconducting layers. The propagation and reflection of the Josephson plasma waves in such systems are viewed in terms of Markov chains. The next application is designing complex diffraction gratings with prescribed characteristics that display simultaneously periodic, quasi-periodic and random gratings properties. Further we study a large class of rank distributions in stochastic systems that obey the power law with the exponent of the order of unity in the case of rather strong persistent correlations. The Zipf law is shown to be valid for the rank distribution of words with lengths about and shorter than the correlation length in the Markov sequence. The last application is devoted to the study of literary texts and DNA "texts". These objects can be viewed as the sequences of random variables. After the appropriate coarse-graining, their statistical properties are easy to study in terms of additive Markov chains. This sheds light upon some information about the author of the text, the style of the text and even a logic of setting out the material of a literary work. The statistical properties of the DNA strings can be used for the study of their different properties. At the end of the Section we touch on the question of the relevance of the coarse graining procedure. It is provided with some simple, but important examples of dynamical systems, which retain their statistical properties after discretization.

Finally, the conclusions based on the prospects and outlooks of further possible developments in the area is made.

2. General theoretical approach

This Section contains the development of the general approach to the additive Markov chains. We begin reviewing the general approach by considering the *binary* N-step Markov chains with the step-wise memory function, see Reference [58].

2.1 MARKOV CHAINS WITH STEP-WISE MEMORY FUNCTIONS

Markov chains with step-wise memory function are very important because they allow us to find many statistical characteristics of the system in explicit analytical form. They are the correlation and distribution functions and the variance of number of symbols in the words of arbitrary length L. This model makes it possible to study numerous applications to real dynamical systems, specifically, literary texts and DNA sequences. The model shed light on a very interesting property, self-similarity of the stochastic chain under study. The corresponding similarity group transformation of the chain parameters is presented. Another feature that was revealed thanks to the explicit expressions for the distribution function of the L-words is the diffusion Fokker-Planck equation governing it. If the persistent correlations are not extremely strong, the distribution function is shown to be the Gaussian with the variance being nonlinearly dependent on L. The theory is generalized to the case of biased chain when $\bar{a} \neq 1/2$, [59].

2.1.1 Formulation of Problem

Let us consider a homogeneous binary sequence of symbols, $a_i = \{0, 1\}$, $i \in \mathbf{Z} = \ldots, -2, -1, 0, 1, 2, \ldots$. To determine the *N-step Markov chain* we have to introduce the *conditional probability* $P(a_i \mid a_{i-N}, a_{i-N+1}, \ldots, a_{i-1})$ of the definite symbol a_i (for example, $a_i = 1$ or $a_i = 0$) occurring after the N-word $T_{N,i}$, where $T_{N,i}$ denotes the sequence of symbols $a_{i-N}, a_{i-N+1}, \ldots, a_{i-1}$. Thus, it is necessary to define 2^N values of the P-function corresponding to each possible configuration of the symbols in the N-word $T_{N,i}$. Since we intend to deal with the sequences possessing the great memory length (in some applications it may be of the order of 10^6, or even more) we need to make some simplifications. We suppose that the P-function has the *additive* form,

$$P(a_i = 1 \mid T_{N,i}) = \sum_{k=1}^{N} f(a_{i-k}, k). \tag{2.1}$$

Here the value $f(a_{i-k}, k)$ is the additive contribution of the symbol a_{i-k} to the conditional probability of the symbol unity occurring at the ith site. Equation (2.1) corresponds to the additive influence of the previous symbols on the generated one. Such a Markov chain is referred to as *additive Markov chain*, [31]. The homogeneity of the Markov chain is provided by the independence of the conditional probability Eq. (2.1) of the index i. It is possible to consider Eq. (2.1) as the first term in expansion of conditional probability in the formal series of terms that correspond to the additive (or unary), binary, ternary, and so on functions up to N-ary (see more details in Section 2.3).

Let us rewrite Eq. (2.1) in an equivalent form,

$$P(a_i = 1 \mid T_{N,i}) = b + \sum_{r=1}^{N} F(r)(a_{i-r} - b), \tag{2.2}$$

with

$$b = \frac{\sum\limits_{r=1}^{N} f(0, r)/N}{1 - \sum\limits_{r=1}^{N} F(r)}, \qquad F(r) = \frac{1}{N}[f(1, r) - f(0, r)]. \tag{2.3}$$

The constant b is the value of a_i averaged over the whole sequence, $b = \bar{a}$. Indeed, according to the ergodicity of the Markov chain, $\bar{a}$ coincides with the value of a_i averaged over the ensemble of realizations of the Markov chain. So, we can write

$$\bar{a} = Pr(a_i = 1) = \sum_{T_{N,i}} P(a_i = 1 \mid T_{N,i}) Pr(T_{N,i}). \tag{2.4}$$

Here $Pr(a_i = 1)$ is the probability of occurring the symbol a_i equal to unity and $Pr(T_{N,i})$ is the probability of occurring the definite word $T_{N,i}$ in the considered ensemble of sequences. Substituting $P(a_i = 1 \mid T_{N,i})$ from Eq. (2.2) into Eq. (2.4) and taking into account the obvious relation $\sum_{T_{N,i}} Pr(T_{N,i}) = 1$, we get,

$$\bar{a} = b - b \sum_{r=1}^{N} F(r) + \sum_{r=1}^{N} F(r) \sum_{T_{N,i}} Pr(T_{N,i}) a_{i-r}. \tag{2.5}$$

The sum $\sum_{T_{N,i}} Pr(T_{N,i}) a_{i-r}$ does not depend on the subscript r and obviously coincides with $\bar{a}$. So, we have $\bar{a} = b + (\bar{a} - b) \sum_{r} F(r)$. From this equation we conclude that $b = \bar{a}$. Thus, we can rewrite Eq. (2.2) as Eq. (1.2) and Eq. (2.3) expresses parameters $\bar{a}$ and $F(r)$ in terms of functions $f(1, r)$ and $f(0, r)$.

It is reasonable to assume the function f to be decreasing with an increase of the distance r between the symbols a_{i-r} and a_i in the Markov chain. However, for the sake of simplicity we consider a step-wise function $f(a_{i-r}, r)$ independent of the second argument r. This means that memory function $F(r)$ is independent of r. As a result, the model is characterized by three parameters only, specifically by $f(0)$, $f(1)$, and N:

$$P(a_i = 1 \mid T_{N,i}) = \frac{1}{N} \sum_{r=1}^{N} f(a_{i-r}). \tag{2.6}$$

Note that the probability P in Eq. (2.6) depends on the numbers of symbols 0 and 1 in the N-word but is independent of the arrangement of the elements a_{i-k}. Instead of two parameters $f(0)$ and $f(1)$ it is convenient to introduce new independent parameters ν and μ (see below Eq. (2.10)),

$$f(0) + f(1) = 1 + 2\nu, \qquad |\nu| < 1/2. \tag{2.7}$$

Parameter ν provides the statistical inequality of the numbers of symbols zero and unity in the Markov chain under consideration. In other words, the chain is biased if $\nu \neq 0$. Indeed, taking into account Eqs. (2.6) and (2.7) and the sequence of equations,

$$P(a_i = 1|T_{N,i}) = \frac{1}{N}\sum_{r=1}^{N} f(\tilde{a}_{i-r}) - 2\nu = P(a_i = 0 \mid \tilde{T}_{N,i}) - 2\nu, \tag{2.8}$$

we can see the lack of symmetry with respect to interchange $\tilde{a}_i \leftrightarrow a_i$ in the Markov chain if $\nu \neq 0$. Here $\tilde{a}_i$ is the symbol "opposite" to a_i, $\tilde{a}_i = 1 - a_i$, and $\tilde{T}_{N,i}$ is the word "opposite" to $T_{N,i}$. Therefore, the probabilities of the words $T_{L,i}$ and $\tilde{T}_{L,i}$ occurring are not equal to each other for any word of the length L. At $L = 1$ this yields nonequal average probabilities that symbols 0 and 1 occur in the chain. Particularly, the probability of symbol 0 occurring is greater by 2ν than that of symbol 1. If $\nu = 0$ we have the non-biased case.

Taking into account the symmetry of the conditional probability P with respect to a permutation of symbols a_i (see Eq. (2.6)), we can simplify the notations and introduce the conditional probability p_k of the symbol zero occurring after the N-word containing k unities, e.g., after the word $(\underbrace{11\ldots1}_{k}\ \underbrace{00\ldots0}_{N-k})$,

$$\begin{aligned} p_k &= P(a_{N+1} = 0 \mid \underbrace{11\ldots1}_{k}\ \underbrace{00\ldots0}_{N-k}) \\ &= \frac{1}{2} + \nu + \mu\left(1 - \frac{2k}{N}\right), \end{aligned} \tag{2.9}$$

with the correlation parameter μ being defined by the relation

$$\mu = \frac{f(0) - f(1)}{2} = f(0) - \frac{1}{2} - \nu. \tag{2.10}$$

We focus mainly our attention on the region of μ determined by the persistence inequality $0 < \mu$. In this case, each of the symbols unity in the preceding N-word promotes the birth of new symbol unity. Nevertheless, the major part of our results is valid for the anti-persistent region $\mu < 0$ as well. Note that inequalities $|\nu| < 1/2$ and $|\mu + \nu| < 1/2$ follow from Eq. (2.9). Without loss of generality, we consider a case $\nu > 0$ only.

A similar rule for the production of an N-word $(a_1, \ldots, a_N)$ that follows after a word $(a_0, a_1, \ldots, a_{N-1})$ was suggested in Reference [5]. However, the conditional probability p_k of the symbols a_N occurring does not depend on the previous ones in the model [5].

In order to investigate the model of step-wise memory function we first study non-biased chains where $\nu = 0$. After this it will be easy to generalize the obtained results to the case of non-biased sequences.

2.1.2 Statistical Properties of non-biased Chains

Main equation. For the stationary Markov chain, the probability $b(a_1a_2\dots a_N)$ of a certain word $(a_1, a_2, \dots, a_N)$ occurring satisfies the condition of compatibility for the Chapman-Kolmogorov equation (see, for example, Reference [60]):

$$b(a_1\dots a_N) = \sum_{a=0,1} b(aa_1\dots a_{N-1})P(a_N \mid a, a_1, \dots, a_{N-1}). \tag{2.11}$$

Thus, we have 2^N homogeneous algebraic equations for the 2^N probabilities b of the N-words occurring and the normalization equation $\sum b = 1$. In the case under consideration, the set of equations can be substantially simplified owing to the following statement.

Proposition 1. *The probability $b(a_1a_2\dots a_N)$ depends on the number k of unities in the N-word only, i. e., it is independent of the arrangement of symbols in the word $(a_1, a_2, \dots, a_N)$.*

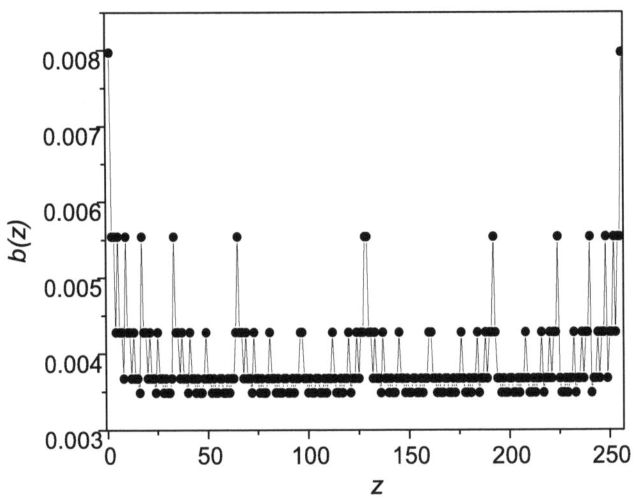

Figure 1 The probability b of occurring a word $(a_1, a_2, \dots, a_N)$ vs its number z expressed in the binary code, $z = \sum_{i=1}^{N} a_i \cdot 2^{i-1}$, for $N = 8$, $\mu = 0.4$.

This statement illustrated by Fig. 1 is valid owing to the chosen simple model (2.9) of the Markov chain. It can be easily verified directly by substituting the solution obtained below (2.15) into the set (2.11). Note that according to the Markov theorem, Eqs. (2.11) do not have other solutions [61].

Proposition 1 leads to the very important property of isotropy: any word $(a_1, a_2, \dots, a_L)$, $L \le N$, appears with the same probability as the inverted one, $(a_L, a_{L-1}, \dots, a_1)$.

Let us apply the set of Eqs. (2.11) to the word $(\underbrace{11\dots1}_{k}\,\underbrace{00\dots0}_{N-k})$:

$$b(\underbrace{11\dots1}_{k}\,\underbrace{00\dots0}_{N-k}) = b(0\,\underbrace{11\dots1}_{k}\,\underbrace{00\dots0}_{N-k-1})p_k + b(1\,\underbrace{11\dots1}_{k}\,\underbrace{00\dots0}_{N-k-1})p_{k+1}. \tag{2.12}$$

This yields the recursion relation for $b(k) = b(\underbrace{11...1}_{k}\ \underbrace{00...0}_{N-k})$,

$$b(k) = \frac{1 - p_{k-1}}{p_k} b(k-1) = \frac{N - 2\mu(N - 2k + 2)}{N + 2\mu(N - 2k)} b(k-1). \tag{2.13}$$

The probabilities $b(k)$ for $\mu > 0$ satisfy the sequence of inequalities,

$$b(0) = b(N) > b(1) = b(N-1) > ... > b(N/2), \tag{2.14}$$

which is the reflection of persistent properties for the chain. At $\mu = 0$ all probabilities are equal to each other.

The solution of Eq. (2.11) is

$$b(k) = A \cdot \Gamma(n+k)\Gamma(n+N-k) \tag{2.15}$$

with the parameter n defined by

$$n = \frac{N(1-2\mu)}{4\mu}. \tag{2.16}$$

The constant A will be found below by normalizing the distribution function. Its value is,

$$A = \frac{4^n}{2\sqrt{\pi}} \frac{\Gamma(1/2+n)}{\Gamma(n)\Gamma(2n+N)}. \tag{2.17}$$

In this Subsection we investigate the statistical properties of the Markov chain, specifically, the distribution of the words of definite length L by the number k of unities. The length L can also be interpreted as the number of jumps of some particle over an integer-valued 1-D lattice or as the time of the diffusion imposed by the Markov chain under consideration. The form of the distribution function $W_L(k)$ depends, to a large extent, on the relation between the word length L and the memory length N. Therefore, the first thing to do is to examine the simplest case $L = N$.

Statistics of N-words. The value $b(k)$ is the probability that an N-word contains k unities with a *definite* order of symbols a_i. Therefore, the probability $W_N(k)$ that an N-word contains k unities with *arbitrary* order of symbols a_i is $b(k)$ multiplied by the number $C_N^k = N!/k!(N-k)!$ of different permutations of k unities in the N-word,

$$W_N(k) = C_N^k b(k). \tag{2.18}$$

Combining Eqs. (2.15) and (2.18), we find the distribution function,

$$W_N(k) = W_N(0) C_N^k \frac{\Gamma(n+k)\Gamma(n+N-k)}{\Gamma(n)\Gamma(n+N)}. \tag{2.19}$$

The normalization constant $W_N(0)$ can be obtained from the equality $\sum_{k=0}^{N} W_N(k) = 1$,

$$W_N(0) = \frac{4^n}{2\sqrt{\pi}} \frac{\Gamma(n+N)\Gamma(1/2+n)}{\Gamma(2n+N)}. \tag{2.20}$$

Comparing Eqs. (2.15), (2.18)–(2.20), we can get Eq. (2.17) for the constant A in Eq. (2.15).

Note that the distribution $W_N(k)$ is an even function of the variable $\kappa = k - N/2$,

$$W_N(N-k) = W_N(k). \tag{2.21}$$

This fact is a direct consequence of the above-mentioned statistical equivalence of zeros and unities in the Markov chain being considered. Let us analyze the distribution function $W_N(k)$ for different relations between the parameters N and μ.

The first case is the limiting case of weak persistence, $n \gg 1$. In terms of the correlation parameter μ, this limiting case corresponds to the values of μ not very close to 1/2,

$$\frac{1-2\mu}{4\mu} \gg \frac{1}{N}. \tag{2.22}$$

This inequality can be rewritten via the f-function (see Eqs. (2.7), (2.10)),

$$\frac{f(1)}{f(0)-f(1)} \gg \frac{1}{N}. \tag{2.23}$$

In the absence of correlations, $n \to \infty$, Eq. (2.19) and the Stirling formula yield the Gaussian distribution at $k,\ N,\ N-k \gg 1$. Given that the persistence is not too strong,

$$n \gg 1, \tag{2.24}$$

we can also obtain the Gaussian form for the distribution function,

$$W_N(k) = \frac{1}{\sqrt{2\pi D(N)}} \exp\left\{-\frac{(k-N/2)^2}{2D(N)}\right\}, \tag{2.25}$$

with the μ-dependent variance,

$$D(N) = \frac{N(N+2n)}{8n} = \frac{N}{4(1-2\mu)}. \tag{2.26}$$

Equation (2.25) says that the N-words with equal numbers of zeros and unities, $k = N/2$, are most probable. Note that the persistence results in an increase of the variance $D(N)$ with respect to its value $N/4$ at $\mu = 0$. In other words, the persistence is conducive to the intensification of the diffusion. Inequality $n \gg 1$ gives $D(N) \ll N^2$. Therefore, despite the increase of $D(N)$, the fluctuations of $(k - N/2)$ of the order of N are exponentially small.

The second case is intermediate one, $n \sim 1$. If the parameter n is an integer of the order of unity, the distribution function $W_N(k)$ is a polynomial of degree $2(n-1)$. In particular, at $n = 1$, the function $W_N(k)$ is constant,

$$W_N(k) = \frac{1}{N+1}. \tag{2.27}$$

At $n \neq 1$, $W_N(k)$ has a maximum in the middle of the interval $[0, N]$.

In the limiting case of strong persistence, when the parameter n satisfies the inequality,

$$n \ll \ln^{-1} N, \tag{2.28}$$

or

$$1-2\mu \ll 1/N\ln(N), \qquad f(1) \ll 1/N\ln(N), \tag{2.29}$$

we can neglect the parameter n in the arguments of the functions $\Gamma(n+k)$, $\Gamma(n+N)$, and $\Gamma(n+N-k)$ in Eq. (2.19). In this case, the distribution function $W_N(k)$ assumes its maximal values at $k=0$ and $k=N$,

$$W_N(1) = W_N(0)\frac{nN}{N-1} \ll W_N(0). \tag{2.30}$$

Formula (2.30) describes the sharply decreasing $W_N(k)$ as k varies from 0 to 1 (and from N to $N-1$). Then, at $1<k<N/2$, the function $W_N(k)$ decreases more slowly with an increase in k,

$$W_N(k) = W_N(0)\frac{nN}{k(N-k)}. \tag{2.31}$$

At $k=N/2$, the probability $W_N(k)$ achieves its minimal value,

$$W_N\left(\frac{N}{2}\right) = W_N(0)\frac{4n}{N}. \tag{2.32}$$

It follows from normalization (2.20) that the values $W_N(0) = W_N(N)$ are approximatively equal to $1/2$. Neglecting the terms of the order of n^2, we get

$$W_N(0) = \frac{1}{2}(1 - n\ln N). \tag{2.33}$$

In the straightforward calculation using Eqs. (1.8) and (2.31) the variance D is

$$D(N) = \frac{N^2}{4} - \frac{nN(N-1)}{2}. \tag{2.34}$$

Thus, the variance $D(N)$ is equal to $N^2/2$ in the leading approximation in the parameter n. This fact has a simple explanation. The probability of occurrence the N-word containing N unities is approximatively equal to $1/2$. So, the relations $\overline{k^2} \approx N^2/2$ and $\overline{k}^2 = N^2/4$ give (2.34). The case of strong persistence corresponds to the so-called ballistic regime of diffusion: if we chose randomly some symbol a_i in the sequence, it will be surrounded by the same symbols with the probability close to unity.

The evolution of the distribution function $W_N(k)$ from the Gaussian form to the inverse one with a decrease of the parameter n is shown in Fig. 2. In the interval $\ln^{-1} N < n < 1$ the curve $W_N(k)$ is concave and the maximum of function $W_N(k)$ inverts into minimum. At $N \gg 1$ and $\ln^{-1} N < n < 1$, the curve remains a smooth function of its argument k as shown by curve with $n=0.5$ in Fig. 2. Here, we will not consider this relatively narrow region of the change in the parameter n.

Formulas (2.25), (2.26), (2.31), (2.33) and (2.34) describe the statistical properties of L-words for the fixed "diffusion time" $L=N$. It is necessary to examine the distribution function $W_L(k)$ for the general situation, $L \neq N$. We start the analysis with $L<N$.

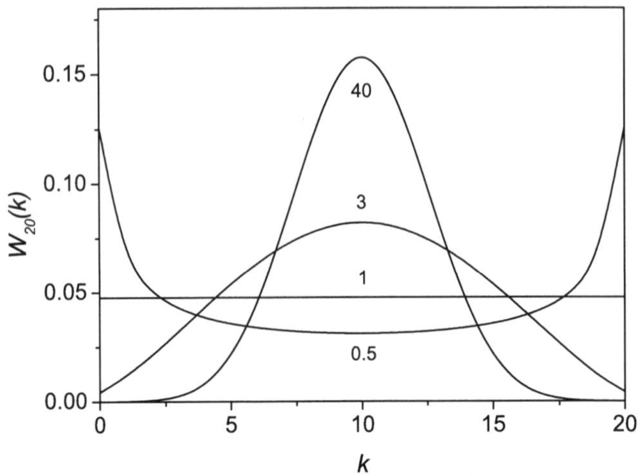

Figure 2 The distribution function $W_N(k)$ for N=20 and different values of the parameter n shown near the curves.

Statistics of L-words with $L < N$. The distribution function $W_L(k)$ at $L < N$ can be given as

$$W_L(k) = \sum_{i=k}^{k+N-L} b(i) C_L^k C_{N-L}^{i-k}. \tag{2.35}$$

This equation follows from the consideration of N-words consisting of two parts,

$$(\underbrace{a_1, \ldots, a_{N-L}}_{i\text{–}k \text{ unities}}, \underbrace{a_{N-L+1}, \ldots, a_N}_{k \text{ unities}}). \tag{2.36}$$

The total number of unities in this word is i. The right-hand part of the word (L-sub-word) contains k unities. The remaining $(i-k)$ unities are situated within the left-hand part of the word (within $(N-L)$-sub-word). The multiplier $C_L^k C_{N-L}^{i-k}$ in Eq. (2.35) takes into account all possible permutations of the symbols "1" within the N-word on condition that the L-sub-word always contains k unities. Then we perform the summation over all possible values of the number i. Note that Eq. (2.35) is a direct consequence of the proposition 1 formulated in the previous paragraph (Statistics of N-words).

The straightforward summation in Eq. (2.35) yields the following formula that is valid at any value of the parameter n:

$$W_L(k) = W_L(0) C_L^k \frac{\Gamma(n+k)\Gamma(n+L-k)}{\Gamma(n)\Gamma(n+L)} \tag{2.37}$$

where

$$W_L(0) = \frac{4^n}{2\sqrt{\pi}} \frac{\Gamma(1/2+n)\Gamma(n+L)}{\Gamma(2n+L)}. \tag{2.38}$$

It is of interest to note that the parameter of persistence μ and the memory length N are presented in Eqs. (2.37), (2.38) via the parameter n only. This means that the statistical properties of the L-words with $L < N$ are defined by this single "combined" parameter.

In the limiting case of weak persistence, $n \gg 1$, at $k,\ L-k \gg 1$, Eq. (2.37) along with the Stirling formula give the Gaussian distribution function,

$$W_L(k) = \frac{1}{\sqrt{2\pi D(L)}} \exp\left\{-\frac{(k-L/2)^2}{2D(L)}\right\} \tag{2.39}$$

with the variance $D(L)$,

$$D(L) = \frac{L}{4}\left(1+\frac{L}{2n}\right) = \frac{L}{4}\left[1+\frac{2\mu L}{N(1-2\mu)}\right]. \tag{2.40}$$

In the case of strong persistence (2.28), the asymptotic expression for the distribution function Eq. (2.37) can be written as

$$W_L(k) = W_L(0)\frac{nL}{k(L-k)}, \quad k \neq 0, \ k \neq L, \tag{2.41}$$

$$W_L(0) = W_L(L) = \frac{1}{2}(1 - n\ln L). \tag{2.42}$$

Both the distribution $W_L(k)$ (2.41) and the function $W_N(k)$ (2.31) have a concave form. The former assumes the maximal value (2.42) at the edges of the interval $[0, L]$ and has a minimum at $k = L/2$.

As previously stated, the important statistical characteristic of the correlated sequence is variance $D(L)$ of the number of unities in the L-word. Using the definition Eq. (1.8) and the distribution function Eq. (2.37) one can obtain a very simple formula for $D(L)$,

$$D(L) = \frac{L}{4}[1+m(L-1)], \tag{2.43}$$

with

$$m = \frac{1}{1+2n} = \frac{2\mu}{N-2\mu(N-1)}. \tag{2.44}$$

Eq. (2.43) shows that the variance $D(L)$ obeys the parabolic law independently of the correlation strength in the Markov chain.

In the case of weak persistence, at $n \gg 1$, we obtain the asymptotics Eq. (2.40). It allows us to analyze the behavior of the variance $D(L)$ with an increase in the "diffusion time" L. At small $mL \ll 1$, the dependence $D(L)$ follows the classical law of the Brownian diffusion, $D(L) \approx L/4$. Then, at $mL \sim 1$, the function $D(L)$ becomes super-linear.

For the case of strong persistence, $n \ll 1$, Eq. (2.43) gives the asymptotic,

$$D(L) = \frac{L^2}{4} - \frac{nL(L-1)}{2}. \tag{2.45}$$

The ballistic regime of diffusion leads to the quadratic law of the $D(L)$ dependence in the zero approximation in the parameter $n \ll 1$.

The unusual behavior of the variance $D(L)$ raises an issue as to what particular type of the diffusion equation corresponds to the nonlinear dependence $D(L)$ in Eq. (2.40). In the following paragraph, when solving this problem, we will obtain the conditional probability $p^{(0)}$ of occurring the symbol zero after a given L-word with $L < N$. The ability to find $p^{(0)}$, with some reduced information about the preceding symbols being available, is very important for the study of the self-similarity of the Markov chain (see the end of this paragraph).

Before this we give the generalization of the diffusion equation at $L < N$, $n \gg 1$. It is quite obvious that the distribution $W_L(k)$ satisfies the equation

$$W_{L+1}(k) = W_L(k)p^{(0)}(k) + W_L(k-1)p^{(1)}(k-1). \tag{2.46}$$

Here $p^{(0)}(k)$ is the probability of occurring "0" after an average-statistical L-word containing k unities and $p^{(1)}(k-1)$ is the probability of occurring "1" after an L-word containing $(k-1)$ unities. At $L < N$, the probability $p^{(0)}(k)$ can be written as

$$p^{(0)}(k) = \frac{1}{W_L(k)} \sum_{i=k}^{k+N-L} p_i b(i) C_L^k C_{N-L}^{i-k}. \tag{2.47}$$

The product $b(i)C_L^k C_{N-L}^{i-k}$ in this formula represents the conditional probability of occurring the N-word containing i unities, the right-hand part of which, the L-sub-word, contains k unities (compare with Eqs. (2.35), (2.36)).

The product $b(i)C_{N-L}^{i-k}$ in Eq. (2.47) is a sharp function of i with a maximum at some point $i = i_0$ whereas p_i obeys the linear law (2.9) (of course, we take into account, that $\nu = 0$). This implies that p_i can be factored out of the summation sign being taken at point $i = i_0$. The asymptotical calculation shows that point i_0 is described by the equation,

$$i_0 = \frac{N}{2} - \frac{L/2}{1 - 2\mu(1 - L/N)} \left(1 - \frac{2k}{L}\right). \tag{2.48}$$

Expression (2.9) taken at point i_0 gives the desired formula for $p^{(0)}$ because

$$\sum_{i=k}^{k+N-L} b(i) C_L^k C_{N-L}^{i-k} \tag{2.49}$$

is obviously equal to $W_L(k)$. Thus, we have

$$p^{(0)}(k) = \frac{1}{2} + \frac{\mu L}{N - 2\mu(N-L)} \left(1 - \frac{2k}{L}\right). \tag{2.50}$$

Let us consider a very important point relating to Eq. (2.48). If the concentration of unities in the right-hand part of the word (2.36) is higher than $1/2$, $k/L > 1/2$, then the most probable concentration $(i_0 - k)/(N - L)$ of unities in the left-hand part of this word is likewise increased, $(i_0 - k)/(N - L) > 1/2$. At the same time, the concentration $(i_0 - k)/(N - L)$ is less than k/L,

$$\frac{1}{2} < \frac{i_0 - k}{N - L} < \frac{k}{L}. \tag{2.51}$$

This implies that the increased concentration of unities in the L-words is necessarily accompanied by the existence of a certain tail with an increased concentration of unities as well. We refer to such a phenomenon as the *macro-persistence*. An analysis performed in the following paragraph (Long-range diffusion) will indicate that the correlation length l_c of this tail is γN with $\gamma \geq 1$ dependent on the parameter μ only. It is evident from the above-mentioned property of the isotropy of the Markov chain that there are two correlation tails from both sides of the L-word.

Note that the distribution $W_L(k)$ is a smooth function of arguments k and L near its maximum in the case of weak persistence and $k,\ L-k \gg 1$. By going over to the continuous limit in Eq. (2.46) and using Eq. (2.50) with the relation $p^{(1)}(k-1) = 1 - p^{(0)}(k-1)$, we obtain the diffusion Fokker-Planck equation for the correlated Markov chain,

$$\frac{\partial W}{\partial L} = \frac{1}{8}\frac{\partial^2 W}{\partial \kappa^2} - \eta(L)\frac{\partial}{\partial \kappa}(\kappa W), \tag{2.52}$$

where $\kappa = k - L/2$ and

$$\eta(L) = \frac{2\mu}{(1-2\mu)N + 2\mu L}. \tag{2.53}$$

Equation (2.52) has a solution of the Gaussian form Eq. (2.39) with the variance $D(L)$ satisfying the ordinary differential equation,

$$\frac{\mathrm{d}D}{\mathrm{d}L} = \frac{1}{4} + 2\eta(L)D. \tag{2.54}$$

Its solution, given the boundary condition $D(0) = 0$, coincides with (2.40).

Now we point to one of the most interesting properties of the Markov chain under consideration, namely, its self-similarity. Let us reduce the N-step Markov sequence by regularly (or randomly) removing some symbols and introduce the decimation parameter λ,

$$\lambda = N^*/N \leq 1. \tag{2.55}$$

Here N^* is a renormalized memory length for the reduced N^*-step Markov chain. According to Eq. (2.50), the conditional probability p_k^* of occurring the symbol zero after k unities among the preceding N^* symbols is described by the formula,

$$p_k^* = \frac{1}{2} + \mu^*\left(1 - \frac{2k}{N^*}\right), \tag{2.56}$$

with

$$\mu^* = \mu\frac{\lambda}{1-2\mu(1-\lambda)}. \tag{2.57}$$

The comparison between Eqs. (2.9) and (2.56) shows that the reduced chain possesses the same statistical properties as the initial one but it is characterized by the renormalized parameters (N^*, μ^*) instead of (N, μ). Thus, Eqs. (2.55) and (2.57) determine the one-parametrical renormalization of the parameters of the stochastic chain defined by Eq. (2.9).

The astonishing property of the reduced sequence consists in that *the variance $D^*(L)$ is invariant with respect to the one-parametric decimation transformation* (2.55), (2.57). In other words, it coincides with the function $D(L)$ for the initial Markov chain:

$$D^*(L) = \frac{L}{4}[1 + m^*(L-1)] = D(L), \qquad L < N^*. \tag{2.58}$$

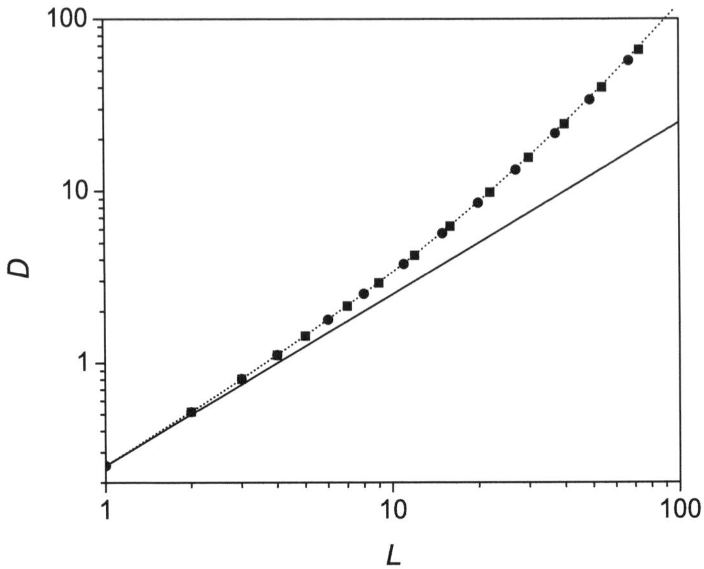

Figure 3 The dependence of the variance D on the word length L for the generated sequence with $N = 100$ and $\mu = 0.4$ (dotted line) and for the decimated sequences (the parameter of decimation $\lambda = 0.5$). Squares and circles correspond to the stochastic and deterministic reduction, respectively. The solid line describes the non-correlated Brownian diffusion, $D(L) = L/4$.

Indeed, according to Eqs. (2.55), (2.57), the renormalized parameter $m^* = 2\mu^*/[N^* - 2\mu^*(N^* - 1)]$ of the reduced sequence coincides exactly with the parameter $m = 2\mu/[N - 2\mu(N - 1)]$ of the initial Markov chain. Since the shape of the function $W_L(k)$ Eq. (2.37) is defined by the invariant parameter $n = n^*$, the distribution $W_L(k)$ is also invariant with respect to the decimation transformation.

The transformation $(N, \mu) \to (N^*, \mu^*)$ (2.55), (2.57) possesses the properties of semi-group, i. e., the composition of transformations $(N, \mu) \to (N^*, \mu^*)$ and $(N^*, \mu^*) \to (N^{**}, \mu^{**})$ with transformation parameters λ_1 and λ_2 is likewise the transformation from the same semi-group, $(N, \mu) \to (N^{**}, \mu^{**})$, with parameter $\lambda = \lambda_1 \lambda_2$.

The invariance of the function $D(L)$ at $L < N$ was referred to by us as the phenomenon of *self-similarity*. It is demonstrated in Fig. 3 and is accordingly discussed below in Long-range diffusion paragraph.

It is interesting to note that the property of self-similarity is valid for any strength of the persistency. Indeed, the result Eq. (2.50) can be obtained directly from Eqs. (2.15)–(2.17), and (2.47) not only for $n \gg 1$ but also for the arbitrary value of n.

Long-range diffusion, $L > N$**.** Unfortunately, the very useful proposition 1 is valid for the words of the length $L \leq N$ only and is not applicable to the analysis of the long words with $L > N$. Therefore, investigating the statistical properties of the long words represents a rather challenging combinatorial problem and requires new physical approaches for its simplification. Thus, we start this paragraph by analyzing the correlation properties of the long words ($L > N$) in the Markov chains with $N \gg 1$. First we mainly deal with the case of relatively weak correlations, $n \gg 1$.

Let us rewrite Eq. (2.9) (with $\nu = 0$) in the form,

$$< a_{i+1} >= \frac{1}{2} + \mu \left(\frac{2}{N} \sum_{k=i-N+1}^{i} < a_k > -1 \right). \quad (2.59)$$

The angle brackets denote the averaging of the density of unities in some region of the Markov chain for its definite realization. The averaging is performed over distances much greater than unity but far less than the memory length N and correlation length l_c (see Eq. (2.64) below). Note that this averaging differs from the statistical averaging over the ensemble of realizations of the Markov chain denoted by the bar,

$$\overline{g(k)} = \sum_{k=0}^{L} g(k) W_L(k). \quad (2.60)$$

Equation (2.59) is a relationship between the average densities of unities in two different macroscopic regions of the Markov chain, namely, in the vicinity of $(i+1)$-th element and in the region $(i-N,\ i)$. Such an approach is similar to the mean field approximation in the theory of the phase transitions and is asymptotically exact at $N \to \infty$. In the continuous limit, Eq. (2.59) can be rewritten in the integral form,

$$< a(i) >= \frac{1}{2} + \mu \left(\frac{2}{N} \int_{i-N}^{i} < a(k) > \mathrm{d}k - 1 \right). \quad (2.61)$$

It has the obvious solution,

$$\left\langle a(i) - \frac{1}{2} \right\rangle = \left\langle a(0) - \frac{1}{2} \right\rangle \exp\left(-i/\gamma N\right), \quad (2.62)$$

where the parameter γ is determined by the relation,

$$\gamma \left(\exp\left(\frac{1}{\gamma}\right) - 1 \right) = \frac{1}{2\mu}. \quad (2.63)$$

A unique solution γ of the last equation is an increasing function of $\mu \in (0,\ 1/2)$.

Formula (2.62) shows that any fluctuation (the difference between $< a(i) >$ and the equilibrium value of $\overline{a_i} = 1/2$) is exponentially damped at distances of the order of the *correlation length* l_c,

$$l_c = \gamma N. \quad (2.64)$$

Law (2.62) describes the phenomenon of the *persistent macroscopic correlations* discussed in the previous paragraph (Statistic of L-words with $L < N$). This phenomenon is governed by both parameters, N and μ. According to Eqs. (2.63), (2.64), the correlation length l_c grows as $\gamma = 1/4\delta$ with an increase in μ (at $\mu \to 1/2$) until the inequality $\delta \gg 1/N$ is satisfied. Here

$$\delta = 1/2 - \mu. \quad (2.65)$$

Let us note that the inequality $\delta \gg 1/N$ defining the regime of weak persistence can be rewritten in terms of γ, $\gamma \ll N/4$. At $\delta \approx 1/N$, the correlation length l_c achieves its maximum value $N^2/4$. With the following increase of μ, the diffusion goes to the regime of strongly correlated diffusion that will be discussed in Subsubsec 3 of this Paragraph.

At $\mu \to 0$, the macro-persistence is broken and the correlation length tends to zero.

Now we study the statistical properties of long words ($L > N$) in the sequences of symbols with strong correlations, $n \ll \ln^{-1} N$. It is convenient to rewrite formula (2.9) for the conditional probability of the symbol zero occurring after the N-word containing k unities in the form,

$$p_{k'} = \delta + 2\mu \frac{k'}{N}, \tag{2.66}$$

where k' is the number of zeros in the precedent N-word, $k' = N - k$.

In the case of strong persistence, $n \ll \ln^{-1} N$, the parameter $\delta = 1/2 - \mu$ is much smaller than $1/N$. Therefore, the probability $p_{k'}$ can be written as

$$p_{k'} \approx \begin{cases} \delta, & k' = 0, \\ k'/N, & k' \neq 0, \quad k' \neq N, \\ 1 - \delta, & k' = N. \end{cases} \tag{2.67}$$

It is seen that the probability of the symbol zero occurring after the N-word which contains only unities ($k' = 0$) represents very small value δ and it increases significantly if $k' \neq 0$. This situation differs drastically from the case of weak persistency. At $n \gg 1$, the parameter δ exceeds the value $1/N$, significantly and the probability p'_k does not actually change with an increase in the number of zeros in the preceding N-word.

The analysis of the symbol generation process in the Markov chain in the case of strong persistence gives the following picture of the fluctuations. There exist the entire portions of the chain consisting of the same symbols, say unities. The characteristic length of such portions is $1/\delta \gg N$ and these portions are separated by one or more zero symbols. The number of such packets of the same symbols in one fluctuation zone is about N. Thus, the characteristic correlation distance at which the N-word containing the same symbols converts into the N-word with $\nu = N/2$ is about N/δ,

$$l_c \approx \frac{N}{\delta}. \tag{2.68}$$

The described structure of the fluctuations defines the statical properties of the L-words with $L > N$ in the case of strong persistence. The distribution function differs significantly from the Gaussian from and is characterized by a concave form at $L \lesssim l_c \sim N/\delta$. As L increases, the correlations between different parts of the L-words get weaker and the L-word can be considered as consisting of a number of independent sub-words. So, according to the general mathematical theorems [64, 65], the distribution function takes on the usual Gaussian form. Such an evolution of the distribution function is depicted in Fig. 4. Note that the Gaussian form of the distribution function is the direct consequence of the finiteness of the correlation length. Thus, the Markov chains studied in this paper differ essentially from strongly correlated systems with $l_c \to \infty$, which are widely discussed in the literature. As is known [see Reference 30], such systems are characterized by the distribution functions with the power-law tails (the so-called Levy distributions).

The variance $D(L)$ follows the quadratic law $D = L^2/4$ (see Eq. (2.45)) up to the range of $L \lesssim l_c \sim N/\delta$ and then approaches to the asymptotics $D(L) = BL$ with $B \sim N/4\delta$ (see Fig. 5).

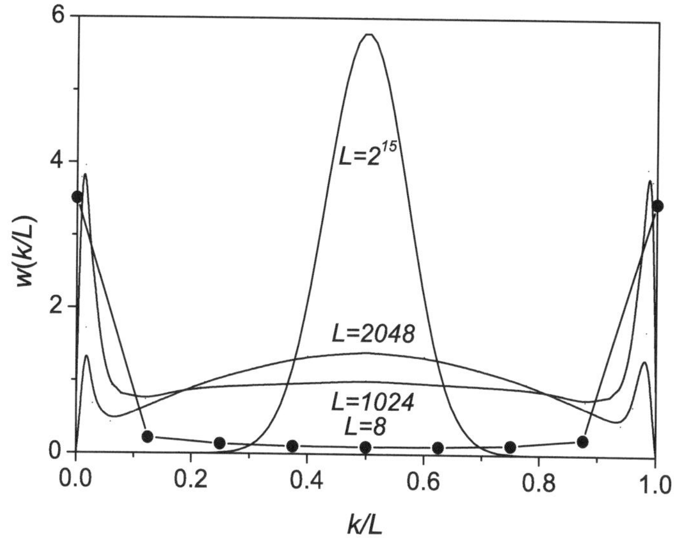

Figure 4 The distribution function $w(k/L) = LW_L(k)$ for N=8 and $\delta = 1/150$. Different values of the length L of words are shown near the curves.

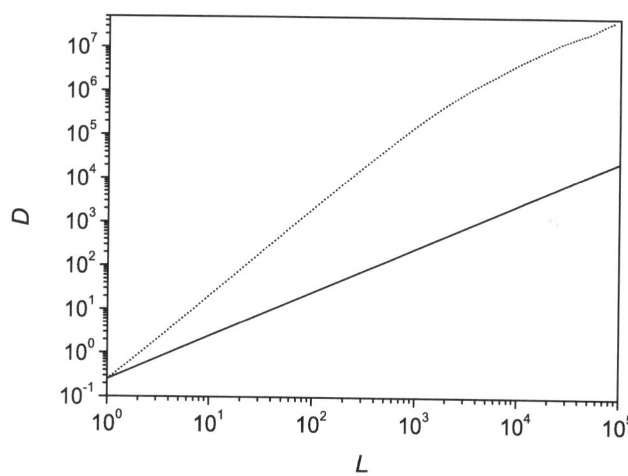

Figure 5 The dependence of the variance D on the word length L for the sequence with $N = 20$ and $\mu = 50/101$ (dotted line). The solid line describes the non-correlated Brownian diffusion, $D(L) = L/4$.

Now we obtain some asymptotical results for the correlation function of the additive Markov chain with an additive memory function. We use the assumption of continuity of correlation function. In Subsection 2.2 we prove this statement to be violated in the general case. However, it is correct in the limit of weak correlations. The results obtained in the following paragraph are used for the analysis of literary text and DNA sequences as it will be shown in Subsection 4.5. The study of the correlation function is carried out more carefully in Subsection 2.2.

Correlation function at weak persistence. Using the studied correlation properties of the Markov sequence and some heuristic reasons, we can obtain the correlation function $K(r)$ and then the variance $D(L)$. Equations (1.10) and (2.43) give the correlation function at $r < N, n \gg 1$,

$$K(r) = \frac{m}{4}, \qquad r \leq N. \tag{2.69}$$

The independence of the correlation function of r at $r < N$ results from our choice of the conditional probability in the simplest form (2.9). At $r > N$, the function $K(r)$ should decrease because of the loss of memory. Therefore, using Eqs. (2.62) and (2.64), let us prolongate the correlator $K(r)$ as the exponentially decreasing function at $r > N$,

$$K(r) = \frac{m}{4} \begin{cases} 1, & r \leq N, \\ \exp\left(-\dfrac{r-N}{l_c}\right), & r > N. \end{cases} \tag{2.70}$$

According to Eqs. (1.11), (2.70), the variance $D(L)$ can be written as

$$D(L) = \frac{L}{4}\,(1 + mF(L)) \tag{2.71}$$

with

$$F(L) = \begin{cases} L, & L < N, \\ 2(1+\gamma)N - (1+2\gamma)\dfrac{N^2}{L} \\ -2\gamma^2 \dfrac{N^2}{L}\left[1 - \exp\left(-\dfrac{L-N}{l_c}\right)\right], & L > N. \end{cases} \tag{2.72}$$

It is clear that the variance (2.71) contains two qualitatively different portions. One of them, at $L \lesssim N$, is the super-linear curve that moves away from the line $D = L/4$ with an increase of L as a result of the persistence. For $L \gg N$, the curve $D(L)$ achieves the linear asymptotics,

$$D(L) \cong \frac{L}{4}\left(1 + \frac{4\mu(1+\gamma)}{1-2\mu}\right). \tag{2.73}$$

This phenomenon can be interpreted as a result of the diffusion in which every *independent* step $\sim \sqrt{D(L)}$ of wandering represents a path traversed by a particle during the characteristic "fluctuating time" $L \sim (N + l_c)$. Since these steps of wandering are quasi-independent, the distribution function $W_L(k)$ is the Gaussian. Thus, in the case of relatively weak persistence, $n \gg 1$, $W_L(k)$ is the Gaussian not only at $L < N$ (see Eq. (2.39)) but also for $L > N,\ l_c$.

Thus we have found explicitly the statistical properties of the sequence in the framework of the discussed model of two-parametrical additive Markov chain. This simplest case is very important because it allows us to calculate such important characteristics of the Markov chain as the distribution function and variance. Now we generalize this model and apply to the case of non-biased chain (with non equal numbers of zeros and unities) that will be characterized by three parameters. All the generalized formulas can be easily obtained and we present only the main results.

2.1.3 Generalization

The idea of finding the distribution function in the case of the biased chain, $\nu \neq 0$, is the same as in the previous Subsection 2.1.2. Here we give this generalization and discuss the main features caused by the non-equal relative numbers of unities and zeros in the chain. The details can be found in Reference [59].

Distribution function of N-words. We start with the recurrence relation (2.13) that in the case of $\nu \neq 1$ is modified,

$$b(k) = \frac{1-p_{k-1}}{p_k} b(k-1) = \frac{N - 2\nu N - 2\mu(N-2k+2)}{N + 2\nu N + 2\mu(N-2k)} b(k-1). \tag{2.74}$$

The solution of Eq. (2.74) is

$$b(k) = A \cdot \Gamma(n_1 + k)\Gamma(n_2 + N - k) \tag{2.75}$$

with the parameters n_1 and n_2 defined by

$$n_1 = \frac{N(1-2(\mu+\nu))}{4\mu}, \quad n_2 = \frac{N(1-2(\mu-\nu))}{4\mu}. \tag{2.76}$$

The constant A will be found below by normalizing the distribution function. Its value is,

$$A = \frac{\Gamma(n_1+n_2)}{\Gamma(n_1)\Gamma(n_2)\Gamma(n_1+n_2+N)}. \tag{2.77}$$

The value $b(k)$ is the probability that an N-word contains k unities with a *definite* order of symbols a_i. Therefore, the probability $W_N(k)$ that an N-word contains k unities with *arbitrary* order of symbols a_i is $b(k)$ multiplied by the number $C_N^k = N!/k!(N-k)!$ of different permutations of k unities in the N-word,

$$W_N(k) = C_N^k b(k). \tag{2.78}$$

Combining Eqs. (2.75) and (2.78), we find the distribution function,

$$W_N(k) = W_N(0) C_N^k \frac{\Gamma(n_1+k)\Gamma(n_2+N-k)}{\Gamma(n_1)\Gamma(n_2+N)}. \tag{2.79}$$

The normalization constant $W_N(0)$ can be obtained from the equality $\sum\limits_{k=0}^{N} W_N(k) = 1$,

$$W_N(0) = \frac{\Gamma(n_1+n_2)\Gamma(n_2+N)}{\Gamma(n_2)\Gamma(n_1+n_2+N)}. \tag{2.80}$$

Comparing Eqs. (2.75), (2.78)–(2.80), we can get Eq. (2.77) for the constant A in Eq. (2.75).

Let us discuss the limiting case of weak persistence, $n_1, n_2 \gg 1$, or

$$\frac{1-2(\mu+\nu)}{4\mu} \gg \frac{1}{N}, \tag{2.81}$$

In the absence of correlations, $n_2 \to \infty$, Eq. (2.79) and the Stirling formula yield the Gaussian distribution at k, Nn_1/n_2, $N-k \gg 1$, $k-k_0 \ll N$. Given the persistence is not too strong,

$$n_2 \gg 1, \tag{2.82}$$

we can also obtain the Gaussian form for the distribution function,

$$W_N(k) = \frac{1}{\sqrt{2\pi D(N)}} \exp\left\{-\frac{(k-k_0)^2}{2D(N)}\right\}, \tag{2.83}$$

with the variance depending on μ and ν,

$$D(N) = \frac{N(N+n_1+n_2)n_1 n_2}{(n_1+n_2)^3} = \frac{N}{4(1-2\mu)}\left[1-\frac{4\nu^2}{(1-2\mu)^2}\right], \tag{2.84}$$

$$k_0 = \frac{n_1}{n_1+n_2}N = \frac{N}{2(1-2\mu)}\left[1-\frac{2\nu}{1-2\mu}\right]. \tag{2.85}$$

It follows from Eq. (2.83) that N-words containing k_0 unties are the most probable. It is interesting to note, that the persistence leads to a *decrease* of the variance $D(N,\, \mu>0)$ with respect to $D(N,\, \mu=0) = N\left(1/4-\nu^2\right)$ if

$$\nu > \frac{1-2\mu}{2\sqrt{3-6\mu+4\mu^2}}. \tag{2.86}$$

In other cases, for instance, at $\nu = 0$, the persistence results in an increase of the variance $D(N,\, \mu)$. To put it differently, the persistence is conducive to the intensification of the diffusion under conditions opposite to inequality (2.86).

Inequality $n_2 \gg 1$ gives $D(N) \ll N^2$. Therefore, despite the increase of $D(N)$, the fluctuations of $(k-k_0)$ of the order of N are exponentially small.

In the intermediate case with the parameters n_1 and n_2 being integers of the order of unity, the distribution function $W_N(k)$ is a polynomial of degree n_1+n_2-2. In particular, at $n_1 = n_2 = 1$, the function $W_N(k)$ is constant,

$$W_N(k) = \frac{1}{N+1}. \tag{2.87}$$

At $n_1 \neq 1$, $W_N(k)$ has a maximum within the interval $[0,\, N]$. At $n_1 = 1$ and $n_2 > 1$, $W_N(k)$ decreases monotonously with an increase of k.

In the case of strong persistence, when the parameter n_2 satisfies the inequality,

$$n_2 \ll \ln^{-1} N, \tag{2.88}$$

we can neglect the parameters n_1 and n_2 in the arguments of the functions $\Gamma(n_1+k)$, $\Gamma(n_2+N)$, and $\Gamma(n_2+N-k)$ in Eq. (2.79). In this case, the distribution function $W_N(k)$ assumes its maximal values at $k=0$ and $k=N$,

$$W_N(1) = W_N(0)\frac{n_1 N}{N-1} \ll W_N(0). \tag{2.89}$$

Formula (2.89) describes the sharply decreasing $W_N(k)$ as k varies from 0 to 1 (and from N to $N-1$). Then, at $1<k<N/2$, the function $W_N(k)$ decreases more slowly with an increase in k,

$$W_N(k) = W_N(0)\frac{n_1 N}{k(N-k)}. \tag{2.90}$$

At $k=N/2$, the probability $W_N(k)$ achieves its minimal value,

$$W_N\left(\frac{N}{2}\right) = W_N(0)\frac{4n_1}{N}. \tag{2.91}$$

It follows from normalization (2.80) that the values $W_N(0)$ and $W_N(N)$ are approximatively equal to $n_2/(n_1+n_2)$ and $n_1/(n_1+n_2)$ respectively. Neglecting the terms of the order of n_2^2, we get

$$W_N(0) = \frac{n_2}{n_1+n_2}(1-n_1\ln N), \tag{2.92}$$

$$W_N(N) = \frac{n_1}{n_1+n_2}(1-n_2\ln N). \tag{2.93}$$

In the straightforward calculation using Eqs. (1.8) and (2.90) the variance D is

$$D(N) = \frac{n_1 n_2 N^2}{(n_1+n_2)^2} - \frac{n_1 n_2 N(N-1)}{n_1+n_2}. \tag{2.94}$$

Thus, the variance $D(N)$ is equal to $n_1 n_2 N^2/(n_1+n_2)$ in the leading approximation in the parameter n. This fact has a simple explanation similarly to the case of non-biased chain. The probability of the occurrence of the N-word containing N unities is approximatively equal to $n_1/(n_1+n_2)$. So, the relations $\overline{k^2} \approx n_1 N^2/(n_1+n_2)$ and $\overline{k}^2 = n_1^2 N^2/(n_1+n_2)^2$ give (2.94).

The evolution of the distribution function $W_N(k)$ from the Gaussian form to the inverse form with a decrease of the parameters n_1 and n_2 is shown in Fig. 6. In the interval $\ln^{-1} N < n_2 < 1$ the curve $W_N(k)$ is concave and the maximum of function $W_N(k)$ inverts into minimum. At $N \gg 1$ and $\ln^{-1} N < n_2 < 1$, the curve remains a smooth function of its argument k as shown by curve with $n=0.5$ in Fig. 6. Below, we will not consider this relatively narrow region of the change in the parameter n_2.

Formulas (2.83), (2.84), (2.90) and (2.92)–(2.94) describe the statistical properties of L-words for the fixed "diffusion time" $L=N$. Now we examine the distribution function $W_L(k)$ for more general situation, $L<N$.

Statistics of L-words with $L<N$. The distribution function $W_L(k)$ at $L<N$ can be given in the case of non-biased chain as:

$$W_L(k) = W_L(0)C_L^k \frac{\Gamma(n_1+k)\Gamma(n_2+L-k)}{\Gamma(n_1)\Gamma(n_2+L)} \tag{2.95}$$

where

$$W_L(0) = \frac{\Gamma(n_1+n_2)\Gamma(n_2+L)}{\Gamma(n_2)\Gamma(n_1+n_2+L)}. \tag{2.96}$$

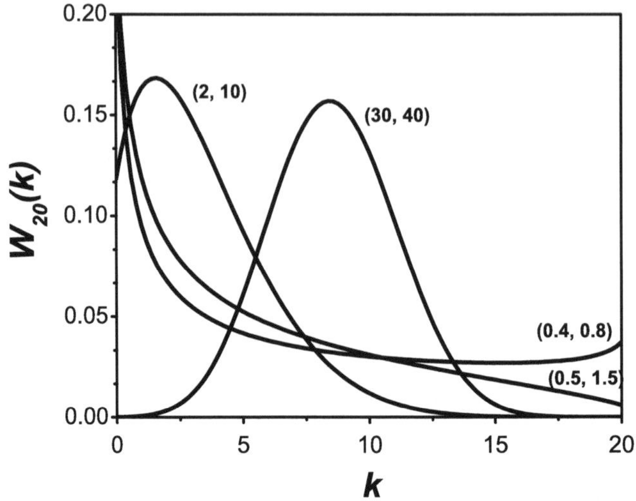

Figure 6 The distribution function $W_N(k)$ for N=20 and different values of the parameters n_1 and n_2 shown near the curves.

Similarly to the case of non-biased chain parameters μ, ν and the memory length N are presented in Eqs. (2.95), (2.96) via the parameters n_1 and n_2 only. This means that the statistical properties of the L-words with $L < N$ are defined by these "combined" parameters.

In the limiting case of weak persistence, $n_2 \gg 1$, at $k,\ Ln_1/n_2,\ L-k \gg 1$, Eq. (2.95) along with the Stirling formula give the Gaussian distribution function,

$$W_L(k) = \frac{1}{\sqrt{2\pi D(L)}} \exp\left\{-\frac{(k-k_0)^2}{2D(L)}\right\} \tag{2.97}$$

with the variance $D(L)$

$$D(L) = \frac{n_1 n_2 L}{(n_1+n_2)^2}\left(1+\frac{L}{n_1+n_2}\right) = \frac{L}{4}\left[1+\frac{2\mu L}{N(1-2\mu)}\right]\left[1-\frac{4\nu^2}{(1-2\mu)^2}\right] \tag{2.98}$$

and

$$k_0 = \frac{n_1 L}{n_1+n_2} = \frac{L}{2}\left[1-\frac{2\nu}{1-2\mu}\right]. \tag{2.99}$$

In the case of strong persistence (2.88), the asymptotic expression for the distribution function Eq. (2.95) can be written as

$$W_L(k) = W_L(0)\frac{n_1 L}{k(L-k)}, \quad k \neq 0,\ k \neq L, \tag{2.100}$$

$$W_L(0) = \frac{n_2}{n_1+n_2}(1-n_1 \ln L), \ W_L(L) = \frac{n_1}{n_1+n_2}(1-n_2 \ln L). \tag{2.101}$$

Both the distribution $W_L(k)$ (2.100) and the function $W_N(k)$ (2.90) have concave forms. The latter yields the maximal values of (2.101) to be at the edges of the interval $[0,\ L]$ and a minimum to be at $k = L/2$.

The definition Eq. (1.8) and the distribution function Eq. (2.95) yields the variance $D(L)$,

$$D(L) = \frac{Ln_1n_2}{(n_1+n_2)^2}\left[1+\frac{(L-1)}{1+n_1+n_2}\right]$$
$$= \frac{L}{4}\left[1+\frac{2\mu(L-1)}{N-2\mu(N-1)}\right]\left[1-\frac{4\nu^2}{(1-2\mu)^2}\right]. \tag{2.102}$$

Eq. (2.102) shows that the variance $D(L)$ obeys the parabolic law independently of the correlation strength in the Markov chain.

In the case of weak persistence, at $n_2 \gg 1$, we obtain the asymptotics Eq. (2.98). It allows us to analyze the behavior of the variance $D(L)$ with an increase in the "diffusion time" L. At small μ, the dependence $D(L)$ follows the classical law of the Brownian diffusion, $D(L) \approx L(1/4-\nu^2)$.

For the case of strong persistence, $n_2 \ll 1$, Eq. (2.102) gives the asymptotics,

$$D(L) = \frac{n_1n_2L^2}{(n_1+n_2)^2} - \frac{n_1n_2L(L-1)}{n_1+n_2}. \tag{2.103}$$

The ballistic regime of diffusion leads to the quadratic law of the $D(L)$ dependence in the zero approximation in the parameter $n_2 \ll 1$.

Using the analogues reasonings to those used in the Subsection 2.1.2 we obtain the diffusion Fokker-Planck equation for the correlated biased Markov chain,

$$\frac{\partial W}{\partial L} = \frac{1}{8}\frac{\partial^2 W}{\partial \kappa^2}\left(1-\frac{4\nu^2}{(1-2\mu)^2}\right) - \frac{2\mu}{(1-2\mu)N+2\mu L}\frac{\partial}{\partial \kappa}(\kappa W), \tag{2.104}$$

where $\kappa = k - L/2$. Equation (2.104) has a solution of the Gaussian form Eq. (2.97).

We finish studying the distribution function of the additive chains with step-wise memory function and proceed to a more general approach. The explicit expressions for the distribution function cannot be obtained by the proposed method even in the case of additive chain with non step-wise memory function. In the next Subsection 2.2 we introduce binary correlation function and derive the recurrence relation for the latter. The obtained relation is applied for the calculation the correlation function of the additive chain with step-wise memory function.

2.2 MEMORY FUNCTION VERSUS BINARY CORRELATOR

In this Subsection we present two proofs of the main relation between the memory function of the additive Markov chain and its correlation function. The first proof follows from a quite natural physical idea while the second one is a bit more complicated but strict from the mathematical point of view. After the equation is obtained we use it to find the correlation function of the chain with the step-wise memory function. Note that this equation can also be employed for finding the memory function of the chain with the definite correlation function and, thus, generating the sequence with the given binary correlation function. This problem will be discussed in Subsection 3.2.

2.2.1 Derivation of Main equation

The physical approach. The first method is based on the minimization of a "distance" *Dist* between the Markov chain generated by means of a sought-for MF and the initial sequence of symbols. This distance is determined by the formula,

$$Dist = \overline{(a_i - P(a_i = 1 \mid T_{N,i}))^2} = \lim_{M\to\infty} \frac{1}{2M+1} \sum_{i=-M}^{M} (a_i - P(a_i = 1 \mid T_{N,i}))^2, \quad (2.105)$$

with the conditional probability P defined by Eq. (1.2).

Let us express distance (2.105) in terms of the correlation function,

$$K(r) = \overline{a_i a_{i+r}} - \bar{a}^2, \quad K(0) = \bar{a}(1-\bar{a}), \quad K(-r) = K(r). \quad (2.106)$$

From Eqs. (1.2), (2.105), one obtains

$$Dist = \sum_{r,r'} \overline{(a_{i-r} - \bar{a})(a_{i-r'} - \bar{a})} F(r)F(r') - 2\sum_{r} \overline{(a_i - \bar{a})(a_{i-r} - \bar{a})} F(r)$$
$$+ \overline{(a_i - \bar{a})^2} = \sum_{r,r'} K(r-r')F(r)F(r') - 2\sum_{r} K(r)F(r) + K(0). \quad (2.107)$$

The minimization equation,

$$\frac{\delta Dist}{\delta F(r)} = 2\sum_{r'} K(r-r')F(r') - 2K(r) = 0, \quad (2.108)$$

yields the relationship between the correlation and memory functions [31],

$$K(r) = \sum_{r'=1}^{N} F(r')K(r-r'), \quad r \geq 1. \quad (2.109)$$

The strict derivation. Equation (2.109) can also be derived by straightforward calculation of the average $\overline{a_i a_{i+r}}$ in Eq. (2.106) using definition (1.2) of the memory function. Let us introduce the function $\phi(r) = p(a_i = 1 \mid a_{i-r} = 1)$, which is the probability of symbol $a_i = 1$ occurring under condition that previous symbol a_{i-r} is also equal to unity. This function is obviously connected to the correlation function $K(r)$, see Eq. (1.6) since the quantity $\overline{a_i a_{i-r}}$ is the probability of simultaneous equality to unity of both symbols, a_i and a_{i-r}. It can be expressed in terms of the conditional probability $\phi(r)$,

$$\overline{a_i a_{i-r}} = \bar{a}\phi(r). \quad (2.110)$$

Substituting Eq. (1.2) into Eq. (1.6) we get:

$$K(r) = \bar{a}\phi(r) - \bar{a}^2. \quad (2.111)$$

For the N-step Markov chain, the probability of the symbol $a_i = 1$ occurring depends only on the previous N-word. Therefore, to obtain the value of $\phi(r)$ we need to average the conditional probability P Eq. (1.2) over all realizations of the N-words with the condition $a_{i-r} = 1$,

$$\phi(r) = p(a_i = 1 \mid a_{i-r} = 1) \tag{2.112}$$
$$= \sum_{T_{N,i}} P(a_i = 1 \mid T_{N,i}) p(T_{N,i} \mid a_{i-r} = 1).$$

If the value of r is less or equal to N, then a_{i-r} in Eq. (2.112) is one of the symbols $a_{i-1}, a_{i-2}, \dots, a_{i-N}$ in the word $T_{N,i}$. In this case, the sum in Eq. (2.112) allows not all N-words, but only the words that contain the symbol unity at the $(i - r)$th position. If $r > N$, the memory function $F(r)$ equals zero in this region and, therefore, the sum in Eq. (2.112) contains *all* terms corresponding to all different N-words.

Substituting Eq. (1.2) into Eq. (2.112), we obtain:

$$\phi(r) = \bar{a} \sum_{T_{N,i}} P(T_{N,i} \mid a_{i-r} = 1) \tag{2.113}$$
$$+ \sum_{r'=1}^{N} F(r') \sum_{T_{N,i}} (a_{i-r'} - \bar{a}) P(T_{N,i} \mid a_{i-r} = 1).$$

According to the normalization condition, the first sum in Eq. (2.113) is equal to unity. Consider the sum

$$\sum_{T_{N,i}} a_{i-r'} P(T_{N,i} \mid a_{i-r} = 1) \tag{2.114}$$

in the second term in RHS of Eq. (2.113). The symbol $a_{i-r'}$ is contained within the word $T_{N,i}$. Therefore, Eq. (2.114) represents the average value of $a_{i-r'}$ under condition $a_{i-r} = 1$. In other words, it equals the probability $\phi(r - r')$ of $a_{i-r'} = 1$ occurring under the condition $a_{i-r} = 1$:

$$\sum_{w} a_{i-r'} P(T_{N,i} \mid a_{i-r} = 1) = \phi(r - r'). \tag{2.115}$$

Substituting this equation into Eq. (2.113) we obtain:

$$\phi(r) = \bar{a} + \sum_{r'=1}^{N} F(r')(\phi(r - r') - \bar{a}). \tag{2.116}$$

Taking into account Eq. (2.111), we arrive at the relation (2.109) connecting the memory function and the correlation function.

Another equation resulting from Eq. (2.109) by double summation over index r establishes a relationship between the memory function $F(r)$ and the variance $D(L)$,

$$M(r, 0) = \sum_{r'=1}^{N} F(r') M(r, r'), \tag{2.117}$$
$$M(r, r') = D(r - r') - (D(-r') + r[D(-r' + 1) - D(-r')]).$$

Equation (1.10) and parity of the function $D(r)$ are used here.

The last equation shows the convenience of using the variance $D(L)$ instead of the correlation function $K(r)$ in numerical calculations. The function $K(r)$, being a second derivative of $D(r)$ in continuous approximation, is less robust in computer simulations. It is the main reason why we prefer to use Eq. (2.117) for analyzing the long-range memory sequences. This is our tool for finding the memory function $F(r)$ of a sequence using the variance $D(L)$, see Subsection 3.2.

2.2.2 Fourier Transformations of Memory and Correlation Functions

In different applications it is more comfortable to operate not with the correlation and memory functions themselves, but with their Fourier transformations. Therefore, we obtain equations for this functions that follow from the main equation (2.109).

For simplicity, we use the binary correlation function $\mathcal{K}(r)$ normalized to the unity at $r = 0$,

$$\mathcal{K}(r) = K(r)/K(0). \tag{2.118}$$

Memory function $F(r)$ is defined for $r \geq 1$, while the correlator $\mathcal{K}(r)$ being an even function, is determined for all values of r. To avoid mathematical problems related to this fact, we assume that the memory function $F(r)$ can be analytically continued to negative values of r, namely, $F(-r) = F(r)$. We have also to specify the value of $F(r)$ for $r = 0$. Further, without loss of generality, we assume that $F(0) = 0$. As a result, we arrive at main equations that shall be analyzed below,

$$\mathcal{K}(r) = \sum_{r'=-\infty}^{\infty} \mathcal{K}(r-r')F(r') - \sum_{r'=1}^{\infty} \mathcal{K}(r+r')F(r'), \qquad r \geq 1; \tag{2.119}$$

$$\mathcal{K}(-r) = \mathcal{K}(r), \quad \mathcal{K}(0) = 1; \tag{2.120}$$

$$F(-r) = F(r), \quad F(0) = 0. \tag{2.121}$$

The above equations can be considered as the starting point for two problems. The first one is to obtain the memory function $F(r)$ by making use of Eq. (2.119) in which the correlator $\mathcal{K}(r)$ is known. The second problem is the complimentary one, namely, to find the correlator $\mathcal{K}(r)$ for a given memory function $F(r)$.

Equation for binary correlator. It is convenient to write the Fourier representation of the main equations (2.119). For this, we introduce the Fourier transform $\hat{\mathcal{K}}(k)$ of the binary correlator $\mathcal{K}(r)$, known as the *randomness power spectrum*,

$$\mathcal{K}(r) = \frac{1}{\pi} \int_0^{\pi} dk\, \hat{\mathcal{K}}(k) \cos(kr), \tag{2.122}$$

$$\hat{\mathcal{K}}(k) = 1 + 2 \sum_{r=1}^{\infty} \mathcal{K}(r) \cos(kr); \tag{2.123}$$

$$\hat{\mathcal{K}}(-k) = \hat{\mathcal{K}}(k), \qquad \hat{\mathcal{K}}(k+2\pi) = \hat{\mathcal{K}}(k). \tag{2.124}$$

Since the correlator $\mathcal{K}(r)$ is a real and even function of the coordinate r, its Fourier transform (2.123) is an even, real and non-negative function of the wave number k.

The condition $\mathcal{K}(r=0)=1$ results in the following normalization relation for the power spectrum $\hat{\mathcal{K}}(k)$,

$$\frac{1}{2\pi}\int_{-\pi}^{\pi} dk\,\hat{\mathcal{K}}(k) = \frac{1}{\pi}\int_{0}^{\pi} dk\,\hat{\mathcal{K}}(k) = 1\,. \tag{2.125}$$

Analogously, the expressions for the Fourier transforms of the memory function $F(r)$ have the form,

$$F(r) = \frac{1}{\pi}\int_{0}^{\pi} dk\,\hat{F}(k)\cos(kr), \tag{2.126}$$

$$\hat{F}(k) = 2\sum_{r=1}^{\infty} F(r)\cos(kr); \tag{2.127}$$

$$\hat{F}(-k) = \hat{F}(k), \qquad \hat{F}(k+2\pi) = \hat{F}(k). \tag{2.128}$$

Similarly, the Fourier transform $\hat{F}(k)$ of the memory function $F(r)$ is also even and real function of the wave number k. Due to the imposed condition $F(r=0)=0$, the normalization condition for $\hat{F}(k)$ reads as,

$$\frac{1}{2\pi}\int_{-\pi}^{\pi} dk\,\hat{F}(k) = \frac{1}{\pi}\int_{0}^{\pi} dk\,\hat{F}(k) = 0. \tag{2.129}$$

The Fourier transform of relation (2.119) (see Appendix B) can be written as the integral equation

$$[\hat{F}(k)-1]\hat{\mathcal{K}}(k) - \frac{1}{2\pi}\int_{-\pi}^{\pi} dk'\,\Upsilon(k,k')\hat{\mathcal{K}}(k') = A-1, \tag{2.130}$$

and the integral equation for the Fourier transform $\hat{F}(k)$ of the memory function,

$$\hat{\mathcal{K}}(k)\hat{F}(k) - \frac{1}{\pi}\int_{0}^{\pi} dk'\,\Xi(k,k')\hat{F}(k') = \hat{\mathcal{K}}(k) + A - 1. \tag{2.131}$$

The obtained equations for the Fourier transformations of functions $\mathcal{K}$ and F are rather complicated. Nevertheless, they can be very important for the analysis of different spectral characteristics of stochastic systems.

2.2.3 Correlation Function of Chains with Step-Wise Memory Functions

In the Subsection 2.2.1, we obtained the relationship (2.109) between two characteristics of the Markov chain, the memory and correlation functions. Here we present the solution of the problem of finding the binary correlation function of the chain with the given memory function,

$$F(r) = \begin{cases} \alpha, & r \leqslant N, \\ 0, & r > N, \end{cases} \tag{2.132}$$

with parameter α restricted by inequality $|\alpha| < 1/N$ (see Eq.(1.2)).

Main equation for the correlation function. Substituting Eq. (2.132) into Eq. (2.109) we obtain the relation,

$$K(r)=\alpha\sum_{r'=1}^{N}K(r-r'),\quad r\geqslant 1. \tag{2.133}$$

Here, the correlation function is assumed to be even, $K(-r)=K(r)$. Equation (2.133) is the linear recurrence of the order of N for $r\geqslant N+1$, so we stand in need of N initial conditions. For the unbiased sequence, $\bar{a}=1/2$, we have $K(0)=1/4$. The solution of Eqs. (2.133) written for $r=1,\dots,N$ yields the constant value of the correlation function, $K(r)=K_0$ at $r=1,\dots,N-1$,

$$K_0=\frac{\alpha}{4(1-\alpha(N-1))}. \tag{2.134}$$

Subtracting Eq. (2.133) from the same equation written for $(r+1)$, we derive another, more convenient, form of the recurrence:

$$K(r+1)-(1+\alpha)\,K(r)+\alpha\,K(r-N)=0. \tag{2.135}$$

This equation is of the order of $N+1$, thus we need an additional initial condition. It can be derived from Eq. (2.133): $K(N)=K_0$. Note that the possibility to rewrite Eq. (2.133) in the form of Eq. (2.135) is the result of the simple structure of the memory function. We solve the obtained recurrent equations by the most natural method, by means of step-by-step finding the sequent values of the correlation function. Such an approach is very suitable for the analysis of correlation function at $r\gtrsim N$.

Correlation function at $r\gtrsim N$.

Band structure of the correlation function. Equation (2.133) allows us to find numerically the unknown correlation function $K(r)$. The result of this step-by-step calculation is presented in Fig. 7 by solid line. We can easily see the discontinuity of $K(r)$ at the point $L=N=100$, the breakpoint of the curve is observed at $L=2N$. This behavior of the correlation function is the result of using the memory function of the step-wise form. To clarify this fact it is convenient to change the variable r by the band number s and the intra-band number ρ:

$$K(r)=K_s(\rho),\qquad r=sN+\rho+1,\quad \rho=0,1,\dots,N-1,\ s=0,1,\dots. \tag{2.136}$$

Within sth band, Eq. (2.135) is the recurrence of the second order with the term $\alpha\,K_{s-1}(\rho)$ that is determined at the previous step, while finding the correlation function for the $(s-1)$th band.

General expression for the correlation function. In the zeroth band ($s=0,\ 1\leqslant r\leqslant N$), as it was shown above, the correlation function is constant,

$$K_0(\rho)=K_0. \tag{2.137}$$

For the first band ($s=1,\ N+1\leqslant r\leqslant 2N$), taking into account that $K(r-N-1)=K_0(\rho)$, we have

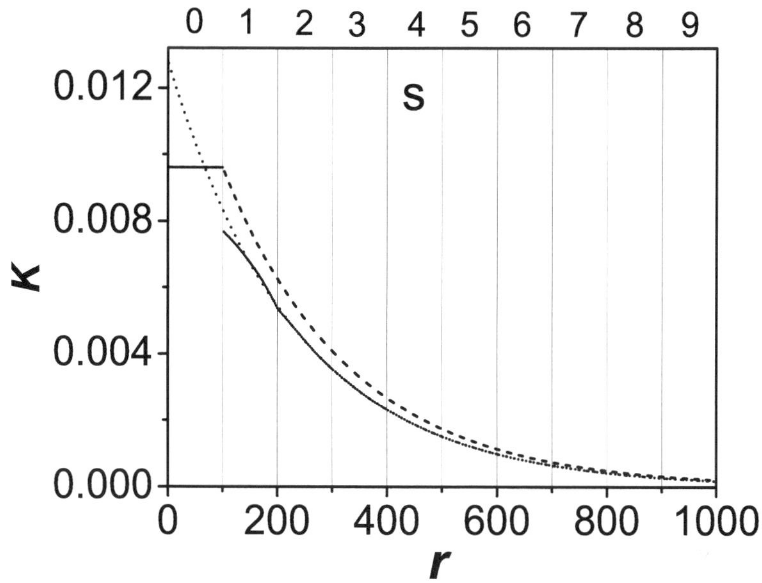

Figure 7 Correlation function $K(r)$ (solid line) obtained by two different methods: the numerical simulation of Eq. (2.109) and exact solution Eq. (2.153). The dotted line is for the contribution to $K(r)$ of the first root of Eq. (2.144). The dashed line refers to the correlation function obtained in [58]. The vertical lines indicate the limits of the bands numbered by s.

$$K_1(\rho) = K_0\,(1-(1-\alpha N)(1+\alpha)^{\rho}). \tag{2.138}$$

The correlation function decreases quasi-continuously within the first band. However, as it was mentioned previously, there exists a discontinuity in the $K(r)$ dependence at $r = N$. This discontinuity disappears in the limiting case of the strong persistence, $\alpha \to 1/N$.

Substituting the obtained formula (2.138) in Eq. (2.135), we find the solution $K_2(\rho)$ for the second band ($s = 2,\ 2N+1 \leqslant r \leqslant 3N$),

$$K_2(\rho) = K_0\,(1-(1-\alpha N)((1+\alpha)^{\rho+N} - \rho\alpha(1+\alpha)^{\rho-1}). \tag{2.139}$$

The correlation function $K(r)$ is continuous at the interface between the first and second bands, $K_1(N) = K_2(0)$. However, its first derivative of $K(r)$ is discontinuous here (see Fig. 7). Using the induction method, we can easily derive the formula for $K_s(\rho)$ in the sth band ($sN+1 \leqslant r \leqslant (s+1)N$):

$$K_s(\rho) = K_0\left(1-(1-\alpha N)\sum_{i=1}^{s}(-\alpha)^{i-1}(1+\alpha)^{(s-i)N+\rho-i+1}C^{i-1}_{(s-i)N+\rho}\right), \tag{2.140}$$

$$C_n^k = \frac{\Gamma(n+1)}{\Gamma(k+1)\Gamma(n-k+1)}.$$

It follows from Eq. (2.140), that the first $(s-2)$ derivatives of the correlation function $K(r)$ are continuous at the border between the $(s-1)$th and sth bands, but the derivative of the $(s-1)$th order changes discontinuously. Under the condition $\alpha N \ll 1$, Eq. (2.140) takes a simpler form,

$$K_s(\rho) = K_0 \alpha^s \mathrm{C}^s_{s+N-1-\rho}. \tag{2.141}$$

It is seen that the correlation function decreases proportionally to α^s with an increase of the band number s.

It is not easy to analyze the asymptotical behavior of the function $K(r)$ at large s because the number of summands in Eq. (2.140) increases being proportional to s. It is the reason for proposing another approach for the asymptotical study of the correlation function $K(r)$ at $s \gg 1$.

Asymptotical study of the correlation function.

Derivation of the characteristic equation. The general solution of linear recursion equations (2.133) can be represented as the linear combination of N different exponential functions,

$$K(r) = \sum_{i=1}^{N} a_i \xi_i^r. \tag{2.142}$$

To find the values of ξ_i, we substitute the fundamental solution,

$$K(r) = \xi^r, \tag{2.143}$$

into Eq. (2.133) and obtain the characteristic polynomial equation of the order of N. Constant multipliers a_i are determined by initial conditions.

It is more convenient to use Eq. (2.135) instead of Eq. (2.133), that implies the characteristic equation of the order of $N+1$,

$$\xi^{N+1} - (1+\alpha)\xi^N + \alpha = 0. \tag{2.144}$$

The extra root of this equation, $\xi = 1$, appears as a consequence of passing on to the equation of order of $N+1$ from that of the order of N . The corresponding coefficient, a_i, in Eq. (2.142) is equal to zero because the correlation function should decrease at $r \to \infty$.

Our study shows that Eq. (2.144) has one real positive root less than unity in the case of odd N. In the case of even N, there are two real roots, one positive and one negative. The remaining roots are complex. All absolute values of roots are less than unity, that is in agreement with the finiteness of memory function $F(r)$. In the case of large N, the absolute magnitudes of all roots are close to unity for nearly all values of α satisfying the inequality,

$$\frac{1}{N} \ln \frac{1}{\alpha} \ll 1. \tag{2.145}$$

Distribution of the roots in the complex plane ξ is shown in Fig. 8.

In the simplest case, $N = 2$, Eq. (2.144) has two real roots,

$$\xi_{1,\,2} = \frac{\alpha}{2} \pm \sqrt{\frac{\alpha^2}{4} + \alpha}. \tag{2.146}$$

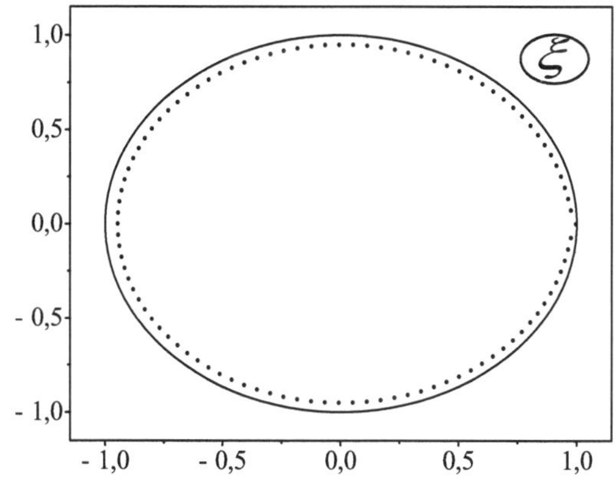

Figure 8 The dots are the roots of the characteristic equation (2.144) for $N=100$ and $\alpha=0.008$. The solid line is the circle $|\xi|=1$.

Taking into account the initial conditions, we find the solution of Eq. (2.133) in the form,

$$K(r)=\frac{\alpha}{4(1-\alpha)\sqrt{\alpha^2+4\alpha}}(\xi_1^{r-1}(1-\xi_2)-\xi_2^{r-1}(1-\xi_1)). \tag{2.147}$$

This expression can be simplified at small and large values of parameter α. For $\alpha \ll 1$, we obtain

$$K(r)=\frac{1}{4}\alpha^{[r/2]+1} \tag{2.148}$$

with square brackets standing for the integer part. The correlation function in the sequent odd and even points are equal to each other. In accordance with Eq. (2.141), $K(r)$ decreases at $r\to\infty$ being proportional to α^s. In the opposite limiting case of the strong persistency, $\alpha\to 1/2$, we have two different roots,

$$\xi_1=1-\frac{4}{3}\phi, \qquad \xi_2=-\frac{1}{2}+\frac{1}{3}\phi, \tag{2.149}$$

with $\phi=1/2-\alpha$. The coefficient corresponding to the second root is much less than that corresponding to the first root. Besides, the second term in Eq. (2.147) decreases more rapidly. Therefore, the approximate solution in this case is

$$K(r)=\frac{1}{4}\exp -4\phi(r-1)/3. \tag{2.150}$$

Correlation function at small α. Let us return to the case of arbitrary value of N. If α is very small, i.e. at

$$\frac{1}{N}\ln\frac{1}{\alpha}\gg 1, \tag{2.151}$$

Eq. (2.144) has N roots with small absolute magnitudes:

$$\xi_k=\alpha^{1/N}\left(\cos\left(2\pi\frac{k}{N}\right)+i\sin\left(2\pi\frac{k}{N}\right)\right), \qquad k=0,\dots,N-1. \tag{2.152}$$

The correlation function, being a linear combination of the power functions with these roots as their exponents, decreases proportionally to α^s, which agrees with Eq. (2.141).

The coefficients a_i in the linear combination Eq. (2.142) can be found in a general case, without any restrictions imposed on the value of α. The solution of Eq. (2.133) written for $1 \leqslant r \leqslant N-1$ along with $K(0)=1/4$ can be expressed with the help of the Vandermond determinants:

$$K(r) = K_0(\alpha N - 1) \sum_{k=1}^{N} \frac{\xi_k^{r-1}}{\prod\limits_{j=1, j\neq k}^{N+1} (\xi_k - \xi_j)}, \tag{2.153}$$

with $\xi_{N+1} = 1$.

Correlation function at not too small α. In the case (2.145) of not too small α, the absolute magnitudes of all roots are close to unity. It is convenient to rewrite Eq. (2.144), introducing two new real variables γ and φ instead of complex x according to

$$x = \left(1 - \frac{1}{\gamma N}\right) e^{i\varphi}. \tag{2.154}$$

Equation (2.144) takes the form,

$$\alpha(e^{\frac{1}{\gamma} - iN\varphi} - 1) = \left(1 - \frac{1}{\gamma N}\right) e^{i\varphi} - 1. \tag{2.155}$$

For the real root, Eq. (2.144) yields,

$$\alpha N \gamma (e^{1/\gamma} - 1) = 1. \tag{2.156}$$

This expression along with Eq. (2.143) determines the asymptotical behavior of correlation function. It was first obtained in Reference [58]. The qualitative approach of this paper did not allow us to take into account the contribution of other roots as it is done in the present study.

Equation (2.155) yields all remaining complex roots with the values of φ, which are quite uniformly distributed over the circle $[0, 2\pi]$ and

$$\frac{1}{\gamma} \sim \ln \frac{1}{\alpha}. \tag{2.157}$$

The roots of Eq. (2.144) located in the vicinity of point $\xi = 1$ are shown in Fig. 9. The single real root is much closer to the line $\mathrm{Re}\ \xi = 1$ than the other ones. Besides, the coefficients a_i in Eq. (2.153) (see also Eq. (2.142)) for all terms containing the complex exponents are much less than those for the term with the real exponent. Therefore, the behavior of correlation function $K(r)$ is determined generally by the term with the real exponent.

The exact correlation function $K(r)$ resulting from the numerical simulation of Eq. (2.153) and its approximation determined by the contribution of the real root alone are shown in Fig. 7. These curves are compared with that obtained in Reference [58] by a qualitative method.

The obtained correlation function can be used to calculate one of the most important characteristics of the random binary sequences, the variance of number of unities in the

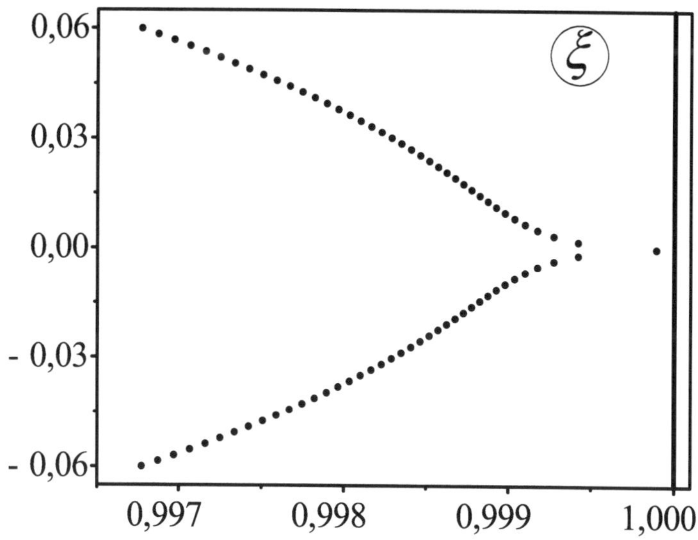

Figure 9 The roots of characteristic polynomial equation close to the point $\xi = 1$ for $N = 4000$, $\alpha = 2 \cdot 10^{-4}$. The solid line is $\mathrm{Re}\ \xi = 1$.

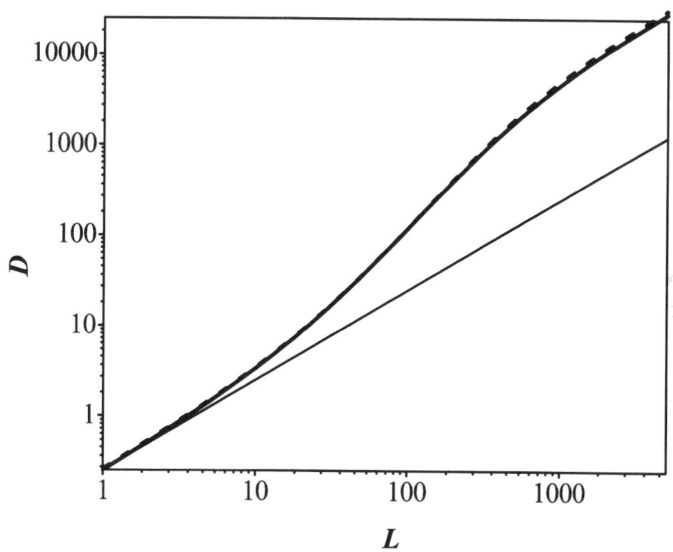

Figure 10 Variance $D(L)$ for the Markov chain with $N = 100$, $\alpha = 0.008$ obtained by means of exact Eqs. (2.153), (1.10) (solid line) and using only one root of Eq. (2.144) (dashed line). The thin solid line describes the non-correlated Brownian diffusion, $D(L) = L/4$.

L-word. The results of the numerical simulations are shown in Fig. 10. We can see a good agreement of curves plotted using both of these methods.

Thus we have expressed the binary correlation function of the additive Markov chain in terms of the memory function and the conditional probability function. For the step-wise memory function the explicit expressions of the binary correlator are found. In different applications we encounter not only with the binary correlation functions but with those of higher orders functions.

2.3 HIGH ORDER CORRELATION FUNCTIONS

The additive Markov chain is the perfect model if we are interested in the binary correlation function of a sequence only. In this case other correlation functions are fixed; they are determined by the memory function. So, we have to study non-additive Markov chains to simulate a sequence with the definite higher correlation functions. In this Subsection we introduce such chains. We present two methods of finding the correlation functions of the high order and illustrate each of them by the applications to the different classes of Markov chains. We consider binary chains only, but some results can be generalized to non-binary sequences as well.

2.3.1 Direct Method of Calculation

In this Subsection, we describe a direct method of finding following characteristics of the N-steps Markov chains: the probability $P(T_N)$ of the N-words occurring, the probability $P(T_L)$ of occurring the words of an arbitrary length L, and the high-order correlation functions.

It is known that all statistical properties of the Markov sequence are determined completely by the conditional probability function. Along with this function, it is convenient to employ the probabilities $P(T_N)$ of the N-words occurring. All other characteristics of the chain can be found using these probabilities. Though the calculation of $P(T_N)$ is quite a challenge in the general case, we succeed in obtaining analytical results for some particular cases (see the third paragraph of this Subsection, Generating the Markov Chain).

Probabilities of the N-words occurring. We start examining the statistical properties of Markov chains with searching the probabilities of the N-words occurring. There are 2^N different N-words in the arbitrary Markov chain which is characterized by 2^N probabilities of these words occurring. They can be found using evident formulas of the probability theory,

$$P(A) = P(A \cap B) + P(A \cap \overline{B}), \tag{2.158}$$

$$P(A \cap B) = P(A|B)P(B). \tag{2.159}$$

Here $A \cap B$ means that events A and B occur simultaneously and event $\overline{B}$ is opposite to the event B. Eqs. (2.158), (2.159) yield the following equation,

$$P(a_1 a_2 \dots a_N) = \sum_{a=0,1} P(a_N | a a_1 \dots a_{N-1}) P(a a_1 \dots a_{N-1}). \tag{2.160}$$

This equation, being written for all possible values of symbols $a_1, \dots, a_N$, along with the normalization requirement,

$$\sum_{\substack{a_i=0,1 \\ i=1,\dots,N}} P(a_1 a_2 \dots a_N) = 1, \tag{2.161}$$

is in fact the set of linear equations. From this set, we can obtain all sought probabilities $P(T_N)$. Though set (2.160), (2.161) contains $2^N + 1$ equations, it has a single solution because some of Eqs. (2.160) are a linear combination of others.

Probabilities of arbitrary words occurring and correlation functions. Using Eq. (2.158) we can calculate the probability $P(T_L)$ with $L < N$ by reducing the N-words to the L-word $(a_1 a_2 \dots a_s)$,

$$P(a_1 a_2 \dots a_L) = \sum_{\substack{a_i = 0,1 \\ i = L+1,\dots,N}} P(a_1 a_2 \dots a_N), \qquad L < N. \tag{2.162}$$

This equation represents the probability the L-word occurring as the average of the probabilities to have the N-word containing this L-word. It is possible to reduce the N-word to the shorter L-word by summing over different sets of symbols at the left and right edges of the N-word.

The word of length L greater than N can be presented as the combination of the word of length N and $(L - N)$ symbols following after it. According to the definition of the Markov chain each of these $(L - N)$ symbols can take on a certain value with the probability depending on the precedent N-word. The $(L - N)$-fold use of Eq. (2.159) yields the following equation,

$$P(a_1 a_2 \dots a_L) = P(a_1 a_2 \dots a_N) \prod_{r=1}^{L-N} P(a_{N+r} | a_r a_{r+1} \dots a_{N+r-1}), \qquad L > N. \tag{2.163}$$

Introducing the notation,

$$R_0 = 0, \qquad R_k = \sum_{i=1}^{k} r_i, \qquad d_i = a_i - \bar{a}, \tag{2.164}$$

we rewrite definition (1.5) of the correlation function,

$$K_s(r_1, r_2, \dots, r_{s-1}) = \sum_{\substack{a_{R_i} = 0,1 \\ i = 0,\dots,s-1}} d_{R_0} d_{R_1} \dots d_{R_{s-1}} P(a_{R_0} a_{R_1} \dots a_{R_{s-1}}). \tag{2.165}$$

Now we can complete the set of symbols $a_{R_0}, a_{R_1}, \dots, a_{R_{s-1}}$ with the symbols between them. Finally we have the following formula,

$$K_s(r_1, r_2, \dots, r_{s-1}) = \sum_{\substack{a_i = 0,1 \\ i = 0,\dots,R_{s-1}}} d_{R_0} d_{R_1} \dots d_{R_{s-1}} P(a_0 a_1 \dots a_{R_{s-1}}). \tag{2.166}$$

Here probabilities $P(a_0 a_1 \dots a_{R_{s-1}})$ should be calculated using Eqs. (2.162) or (2.163).

Generating the Markov chain. Let us apply the proposed method for clarifying generating the Markov chain with a given conditional probability function. We consider generating of the chain with a finite length. Thus the problem is how to generate the first N-word of the chain. Two different simulating procedures can be suggested. In both of them, we generate sequentially each symbol a_i, for $i > N$, using known values of N as previously and the probability function Eq. (1.1). However, the suggested procedures differ in the method of constructing the first N-word, i.e., the initial N symbols.

The first approach is based on the random choice of the initial N-word. For the second approach we need to calculate some additional conditional probabilities

$P_k(a_i = a|a_{i-k}, \dots, a_{i-1})$ of the symbol $a_i = a$ occurring after the k-word $a_{i-k}, \dots, a_{i-1}$, $k < N$. They can be found from the compatibility equation for the conditional probabilities:

$$P_k(a_i = a|a_{i-k}, \dots, a_{i-1}) = \frac{\sum\limits_{a_{i-N}} \dots \sum\limits_{a_{i-k-1}} P_N(a_i = a|a_{i-N}, \dots, a_{i-1}) P(a_{i-N}, \dots, a_{i-1})}{\sum\limits_{a_{i-N}} \dots \sum\limits_{a_{i-k-1}} P(a_{i-N}, \dots, a_{i-1})}. \tag{2.167}$$

Here the sign $\sum\limits_{a_j}$ denotes the summation (or integration) over all possible values of symbol a_j. The probabilities of N-words occurring, $P(a_{i-N}, \dots, a_{i-1})$, should be obtained from the following linear system,

$$\begin{cases} P(a_1, a_2, \dots, a_N) = \sum\limits_{a_0} P_N(a_N|a_0, a_1, \dots, a_{N-1}) P(a_0, a_1, \dots, a_{N-1}), \\ \sum\limits_{a_1} \dots \sum\limits_{a_N} P(a_1, a_2, \dots, a_N) = 1. \end{cases} \tag{2.168}$$

Using Eq. (2.167), we construct the first N symbols sequentially generating them by means of the conditional probabilities:

$$P_0(a_1),\ P_1(a_2|a_1),\ P_2(a_3|a_1, a_2), \dots, P_{N-1}(a_N|a_1, \dots, a_{N-1}).$$

The first approach is simpler than the second one because it does not require that additional probabilities be calculated. However, it does not allow us to get the stationary chain, as it is possible to do using the second method. To generate the same chain using the first approach we must construct so many symbols as we need to obtain the stationary chain (the initial non-stationary part of the chain should be removed).

Necessary conditions for correlation function existence. Here we formulate some important necessary conditions using the direct method. They allow us to analyse possible values of the binary correlation functions of the binary chains. With reference to constructing the sequences with a given binary correlation function, it is of great importance to know what are the restrictions on the type of the correlation functions allowed for binary sequences. To shed light on this problem, here we consider a statistically homogeneous random sequence of symbols s_n consisting of the values "-1" and "1",

$$s_n = \{-1, 1\}, \qquad n \in \mathbf{Z} = \dots, -2, -1, 0, 1, 2, \dots \tag{2.169}$$

The canonical definition of the correlation function reads as follows

$$K_s(r) \equiv \overline{s_n s_{n+r}} - \bar{s}^2 = K_s(0)\mathcal{K}_s(r), \tag{2.170}$$

where $\bar{s} \equiv \overline{s_n}$ and $K_s(0) \equiv \overline{s_n^2} - \bar{s}^2$ are the mean value and variance of s_n, respectively. Taking into account a peculiar property

$$\overline{s_n^2} = s_n^2 = 1 \tag{2.171}$$

of our binary sequence s_n, we obtain

$$K_s(0) = 1 - \bar{s}^2. \tag{2.172}$$

It should be emphasized that the direct relation between the mean value and higher one-point moments is a specific property of *any* binary chain. For instance, for the binary chain ε_n consisting of "0" and "1" we have the relation $K_\varepsilon(0)=\overline{\varepsilon}(1-\overline{\varepsilon})$ since $\varepsilon_n^2=\varepsilon_n$ in this case. On the contrary, for a sequence of continuous random numbers of the Gaussian type the variance and mean value are independent parameters.

To proceed with the two-point moments, we associate the correlator $K_s(r)$ with the probabilities of two symbols with the same or opposite signs, occurring at the distance r,

$$4P(\pm 1, \underbrace{\ldots}_{r-1}, \pm 1) = (\overline{s}\pm 1)^2 + K_s(r), \tag{2.173a}$$

$$4P(\pm 1, \underbrace{\ldots}_{r-1}, \mp 1) = 1-\overline{s}^2 - K_s(r). \tag{2.173b}$$

These relationships are also particular properties solely of binary sequences, they are drastically distinct from the analogous relations for other random processes (see, e.g., Eq. (D.8) for the two-point probability density of the Gaussian process). This fact is strictly confirmed by straightforward calculations of Eqs. (2.173).

In order to obtain the expression (2.173a), we should write the average $\overline{(s_n\pm 1)(s_{n+r}\pm 1)}$ via the correlator (2.170),

$$\overline{(s_n\pm 1)(s_{n+r}\pm 1)}=\overline{s_n s_{n+r}}\pm\overline{s_n}\pm\overline{s_{n+r}}+1=K_s(r)+(\overline{s}\pm 1)^2. \tag{2.174}$$

On the other hand, the same average can be calculated with the use of two-symbol probabilities,

$$\overline{(s_n+1)(s_{n+r}+1)}=2\cdot 2\cdot P(1,\underbrace{\ldots}_{r-1},1)+0\cdot 2\cdot P(-1,\underbrace{\ldots}_{r-1},1)$$
$$+2\cdot 0\cdot P(1,\underbrace{\ldots}_{r-1},-1)+0\cdot 0\cdot P(-1,\underbrace{\ldots}_{r-1},-1)=4P(1,\underbrace{\ldots}_{r-1},1). \tag{2.175}$$

$$\overline{(s_n-1)(s_{n+r}-1)}=0\cdot 0\cdot P(1,\underbrace{\ldots}_{r-1},1)+(-2)\cdot 0\cdot P(-1,\underbrace{\ldots}_{r-1},1)$$
$$+0\cdot(-2)\cdot P(1,\underbrace{\ldots}_{r-1},-1)+(-2)\cdot(-2)\cdot P(-1,\underbrace{\ldots}_{r-1},-1)=4P(-1,\underbrace{\ldots}_{r-1},-1) \tag{2.176}$$

Then, the combination of Eqs. (2.174), (2.175), and (2.176) results in the equality (2.173a).

Similarly, in order to derive the expression (2.173b), we can write,

$$\overline{(s_n\pm 1)(s_{n+r}\mp 1)}=\overline{s_n s_{n+r}}\mp\overline{s_n}\pm\overline{s_{n+r}}-1=K_s(r)+\overline{s}^2-1. \tag{2.177}$$

Again, the above average can be calculated employing two-symbol probabilities. Specifically,

$$\overline{(s_n+1)(s_{n+r}-1)}=2\cdot 0\cdot P(1,\underbrace{\ldots}_{r-1},1)+0\cdot 0\cdot P(-1,\underbrace{\ldots}_{r-1},1)$$
$$+2\cdot(-2)\cdot P(1,\underbrace{\ldots}_{r-1},-1)+0\cdot(-2)\cdot P(-1,\underbrace{\ldots}_{r-1},-1)=-4P(1,\underbrace{\ldots}_{r-1},-1). \tag{2.178}$$

$$\overline{(s_n-1)(s_{n+r}+1)}=0\cdot 2\cdot P(1,\underbrace{\ldots}_{r-1},1)+(-2)\cdot 2\cdot P(-1,\underbrace{\ldots}_{r-1},1)$$
$$+0\cdot 0\cdot P(1,\underbrace{\ldots}_{r-1},-1)+(-2)\cdot 0\cdot P(-1,\underbrace{\ldots}_{r-1},-1)=-4P(-1,\underbrace{\ldots}_{r-1},1). \quad (2.179)$$

From Eqs. (2.177), (2.178), and (2.179) it follows the equality (2.173b).

Now, with the use of expressions (2.173a) we express the correlation function $K_s(r)$ via the probabilities to occur three symbols,

$$K_s(r)+(\bar{s}\pm 1)^2=4\sum_{a=-1,1} P(\pm 1,\underbrace{\ldots}_{r'-1},a,\underbrace{\ldots}_{r-r'-1},\pm 1). \quad (2.180)$$

Probability $P(\pm 1,\underbrace{\ldots}_{r'-1},a,\underbrace{\ldots}_{r-r'-1},\pm 1)$ is smaller than both probabilities $P(\pm 1,\underbrace{\ldots}_{r'-1},a)$ and $P(a,\underbrace{\ldots}_{r-r'-1},\pm 1)$. Thus, we can write,

$$\begin{aligned}K_s(r)+(\bar{s}\pm 1)^2 &\leqslant 4\sum_{a=-1,1}\min\{P(\pm 1,\underbrace{\ldots}_{r'-1},a),\,P(a,\underbrace{\ldots}_{r-r'-1},\pm 1)\}\\ &=\min\{1-\bar{s}^2-K_s(r'),\,1-\bar{s}^2-K_s(r-r')\}\\ &\quad+\min\{(\bar{s}\pm 1)^2+K_s(r'),\,(\bar{s}\pm 1)^2+K_s(r-r')\}\\ &=1-\bar{s}^2+\min\{-K_s(r'),\,-K_s(r-r')\}\\ &\quad+(\bar{s}\pm 1)^2+\min\{K_s(r'),\,K_s(r-r')\}.\end{aligned} \quad (2.181)$$

Here we again have used Eqs. (2.173). Then, according to the evident relation

$$\min\{x,\,y\}+\min\{-x,\,-y\}=-|x-y|, \quad (2.182)$$

we arrive at the condition

$$|K_s(r')-K_s(r-r')|+K_s(r)\leqslant 1-\bar{s}^2. \quad (2.183)$$

Finally, it is convenient to rewrite Eq. (2.183) for the normalized correlator $\mathcal{K}_s(r)$,

$$|\mathcal{K}_s(r')-\mathcal{K}_s(r-r')|+\mathcal{K}_s(r)\leqslant 1. \quad (2.184)$$

Although we have derived this inequality for $0<r'<r$, a simple analysis reveals its validity for arbitrary values of r and r'. We would like to stress that the condition (2.184) is applicable for any mean value $\bar{s}$ of the binary sequence s_n. However, without a loss of generality in what follows we consider binary sequences with the zero mean, since the statistical properties of considered sequences do not depend on mean values.

In a similar manner we can obtain second inequality with the use of Eq. (2.173b),

$$\begin{aligned}1-\bar{s}^2-K_s(r) &= 4\sum_{a=-1,1} P(\pm 1,\underbrace{\ldots}_{r'-1},a,\underbrace{\ldots}_{r-r'-1},\mp 1)\\ &\leqslant 4\sum_{a=-1,1}\min\{P(\pm 1\underbrace{\ldots}_{r'-1},a),\,P(a,\underbrace{\ldots}_{r-r'-1},\mp 1)\}\\ &=\min\{1-\bar{s}^2-K_s(r'),\,(\bar{s}\mp 1)^2+K_s(r-r')\}\\ &\quad+\min\{(\bar{s}\pm 1)^2+K_s(r'),\,1-\bar{s}^2-K_s(r-r')\}.\end{aligned} \quad (2.185)$$

Since $K_s(r) = \mathcal{K}_s(r)$ in the case of $\overline{s} = 0$, Eq. (2.185) gets a simpler form,

$$|\mathcal{K}_s(r') + \mathcal{K}_s(r - r')| - \mathcal{K}_s(r) \leqslant 1 \quad \text{for} \quad \overline{s} = 0. \tag{2.186}$$

Thus, applying the necessary conditions (2.184) and (2.185), or (2.186) if $\overline{s}$=0, we can identify the functions that *can not be* treated as binary correlators of a binary sequence.

Inequalities (2.184), (2.185) and (2.186) have to be met for any values of indices r and r' and, therefore, they actually represent an infinite set of necessary conditions. Evidently, in every particular case we should choose the strongest condition. On the other hand, Eqs. (2.184), (2.185) and (2.186) are automatically satisfied if one of the indices equals zero, $r = 0$ or $r' = 0$. The same takes place when $r = r'$. Summarizing all these facts, we can combine Eqs. (2.184) and (2.186) as follows

$$\begin{aligned} &\max\left\{|\mathcal{K}_s(r') \pm \mathcal{K}_s(r - r')| \mp \mathcal{K}_s(r)\right\} \leqslant 1, \\ &r \neq 0, \quad r' \neq 0, \quad r \neq r' \qquad \text{for} \quad \overline{s} = 0. \end{aligned} \tag{2.187}$$

The symbol $\max\{\dots\}$ implies the absolute maximum of a function with respect to the indices r, r'. Note that Eq. (2.187) is automatically fulfilled for two limit cases, namely, for the delta-correlated (white noise) chain, and for the sequences with infinitely long-range correlations. As one can see, for both these cases $\mathcal{K}_s(r) = \delta_{r,0}$ or $\mathcal{K}_s(r) = 1$, respectively.

2.3.2 Permutative Markov Chains

Solving linear system (2.160), (2.161) for the general case is an immensely complicated problem. Here we demonstrate the application of the described method to some classes of Markov chains. We suppose that the conditional probability function of the chain under consideration is independent of the order of symbols in the previous N-word. We refer to such sequences as the *permutative* Markov chains.

Probabilities of word in the permutative Markov chain. It is convenient to introduce a new shorter notation for the conditional probability function $P(a_{N+1}|a_1a_2\dots a_N) = p_k(a_{N+1})$. Here $k = a_1 + a_2 + \dots + a_N$ is the number of unities in N-word $(a_1a_2\dots a_N)$. Besides, we define $p_k(1)$ as p_k.

Let us seek the solution of system (2.160), (2.161) in the form

$$b_N(k) = b_N(a_1 + a_2 + \dots + a_N) = P(a_1a_2\dots a_N). \tag{2.188}$$

In other words, the probability of the N-word occurring depends on the number of unities in this N-word only. Then Eq. (2.160) can be rewritten as the following recurrence relation,

$$b_N(k) = p_{k-1}b_N(k-1) + p_kb_N(k). \tag{2.189}$$

The solution of this equation is

$$b_N(k) = b_N(0)\prod_{r=1}^{k}\frac{p_{r-1}}{1 - p_r}. \tag{2.190}$$

Here probability $b_N(0)$ can be obtained from Eq. (2.161),

$$b_N(0) = \left(\sum_{k=0}^{N} C_N^k \prod_{r=1}^{k} \frac{p_{r-1}}{1-p_r} \right)^{-1}, \tag{2.191}$$

$$C_n^k = \frac{n!}{k!(n-k)!} = \frac{\Gamma(n+1)}{\Gamma(k+1)\Gamma(n-k+1)}.$$

The probability $P(a_1 a_2 \ldots a_N)$, Eq. (2.188), does not depend on the order of symbols in N-word $(a_1 a_2 \ldots a_N)$. It follows from Eq. (2.162) and this independence that the probability $P(T_L)$ of a short word (with $L < N$) occurring is also independent of the order of symbols in this word. Denoting this probability by $b_L(k) = b_L(a_1 + a_2 + \ldots + a_L) = P(a_1 a_2 \ldots a_L)$ we obtain the following formula,

$$b_L(k) = \sum_{m=0}^{N-L} C_{N-L}^m b_N(k+m). \tag{2.192}$$

The probability of the long L-word (with $L > N$) occurring **does depend** on the order of the symbols in this word. Using Eq. (2.163) we obtain

$$P(a_1 a_2 \ldots a_L) = b_N(q_0) \prod_{r=1}^{L-N} p_{q_{r-1}}(a_{N+r}). \tag{2.193}$$

Here $q_i = a_{i+1} + a_{i+2} + \ldots + a_{i+N}$. Equations (2.190)-(2.192) are generalization of Eqs. (2.15),(2.75) obtained in the Subsection 2.1 to the additive binary Markov chain with the step-wise memory function.

Correlation functions. Now we present analytical results of calculation of the correlation functions of arguments $r_1, \ldots, r_{s-1}$ satisfying the condition $r_1 + \ldots + r_{s-1} < N$. For this purpose we express the average number of unities in the chain using Eq. (2.192),

$$\bar{a} = b_1(1) = \sum_{k=1}^{N} b_N(k) C_{N-1}^{k-1}. \tag{2.194}$$

For the permutative Markov chains under consideration, the correlation function of arguments $r_1, \ldots, r_{s-1}$ depends on their number s only. Equation (2.166) yields

$$K_s(r_1, r_2, \ldots, r_{s-1}) = K_s = \sum_{k=0}^{N} b_N(k) S_N(k, s, \bar{a}), \tag{2.195}$$

$$S_N(k, s, \bar{a}) = \sum_{j=0}^{\min\{s,k\}} (-\bar{a})^{s-j} C_{N-j}^{k-j} C_s^j. \tag{2.196}$$

In particular, the binary correlation function $K_2(r)$ of any permutative chain is constant for the values of arguments, less than memory depth: $K_2(r) = K_2, \quad r < N$.

If we apply all obtained formulas to the additive Markov chain with the step-wise memory function,

$$p_k = \frac{1}{2} - \nu + \mu\left(\frac{2k}{N} - 1\right), \tag{2.197}$$

we get the correlation functions of order s,

$$K_s = \frac{\Gamma(n_1+n_2)}{\Gamma(n_1)} \sum_{k=0}^{s} (-\bar{a})^{s-k} C_s^k \frac{\Gamma(n_1+k)}{\Gamma(n_1+n_2+k)}. \tag{2.198}$$

Here

$$\bar{a} = \frac{n_1}{n_1+n_2}, \quad n_1 = \frac{N(1-2(\mu+\nu))}{4\mu}, \quad n_2 = \frac{N(1-2(\mu-\nu))}{4\mu}. \tag{2.199}$$

For $s=2$ we recover Eq. (2.134) obtained in Subsection 2.2 for the binary correlation function K_2.

Self-similarity of the permutative Markov chain. Here we point out an interesting property of the permutative Markov chains, namely, their self-similarity. This issue was discussed in Subsection 2.1.2 for the step-wise memory function.

Let us reduce the N-step Markov sequence by regularly (or randomly) removing some symbols and introduce the decimation parameter $\lambda = N^*/N$, $1/N < \lambda < 1$, which represents the fraction of symbols kept in the chain.

Due to approachment of symbols in the sequence after the decimation procedure, the binary correlation function $K_2(r)$ transforms into

$$K_2^*(r) = \begin{cases} (1-\{r/\lambda\})K_2([r/\lambda]) + \{r/\lambda\}K_2([r/\lambda]+1), & \text{regular decimation,} \\ \sum\limits_{l=0}^{\infty} K_2(r+l)C_{l+r-1}^{l}(1-\lambda)^l \lambda^r, & \text{random decimation.} \end{cases} \tag{2.200}$$

Here $[f]$ is the entire and $\{f\}$ is the fractional part of f.

The correlation function $K_2(r)$ of the permutative Markov chain is equal to constant value K_2 at arguments $r \leqslant N-1$. According to Eq. (2.200) (regular decimation), the values of the correlation function $K_2^*(r)$ at $r \leqslant \lambda(N-1)$ take on the same value K_2. In the case of the random decimation we have only exponentially small (at $N \gg 1$) differences between the correlation function $K_2^*(r)$ and value K_2,

$$|K_2^*(r) - K_2| \leqslant \frac{A}{\sqrt{\bar{\lambda}N}} \exp(-a(\bar{\lambda}, \lambda)N), \qquad r = \bar{\lambda}N, \qquad \bar{\lambda} < \lambda. \tag{2.201}$$

Here new function $a(\bar{\lambda}, \lambda)$ and constant A are introduced,

$$a(\bar{\lambda}, \lambda) = \bar{\lambda} \ln\left(\frac{\bar{\lambda}}{\lambda}\right) + (1-\bar{\lambda}) \ln\left(\frac{1-\bar{\lambda}}{1-\lambda}\right) > 0, \qquad A = \frac{e}{\sqrt{\pi}}(|K_2| + \max_{r \geqslant N} |K_2(r)|). \tag{2.202}$$

Thus, the correlation function of the decimated sequence is constant (or asymptotically constant) in the region $r < N^*$. This property was referred to as the *self-similarity*.

Self-similarity is the intrinsic property of the permutative Markov chains. It is possible to show that, in the general case, the conditional probability function of decimated N-step Markov sequence is of infinite memory length. This is because its conditional probability is "blurred" by the decimation procedure and becomes to be dependent on **all** previous symbols.

So, self-similarity is the property of the binary correlation function only and is typical of all random sequences with a constant binary correlation function $K_2(r)$ at $r < N$,

$$K_2(r) = K_2,\ 1 \leqslant r < N \Rightarrow \text{Self-similarity.} \tag{2.203}$$

2.3.3 Correlation Functions and Characteristic equations

In this Subsection, we obtain recurrence relations for finding the correlation functions. In the first paragraph we show that these relations yield the explicit expressions for the correlation functions via the root of the characteristic equations. The second paragraph (Application of the algorithm) is devoted to applications of this method. The last paragraph (Generalized algorithm of finding the correlation functions) contains some generalizations of the above-mentioned recurrence relations.

The procedure of finding the correlation functions K_s is based on the mathematical induction method. In other words, we suppose that all correlation functions K_l of the orders less than s are found. For the convenience sake, we admit that $K_0 = 1$ and $K_1 = 0$.

Correlation functions of high orders for the additive Markov chain. Consider the N-step Markov chain with the *additive* conditional probability function Eq. (1.2).

Let us find the recurrence relations for the correlation functions of N-step additive Markov chain. For this purpose, first of all we calculate explicitly the average over symbol $a_{r_1+\cdots+r_{s-1}}$ in Eq. (1.5). Using the notation of Section 2.3.1 and Eq. (1.2), and taking into account equation $P(a_{R_{s-1}} = 1|\cdot) + P(a_{R_{s-1}} = 0|\cdot) = 1$, we can rewrite Eq. (1.5) for arbitrary $r_i > 0$, $i = 1, \ldots, s-1$, in the form,

$$\overline{(a_{R_0} - \bar{a}) \ldots (a_{R_{s-1}} - \bar{a})} =$$

$$\overline{(a_0 - \bar{a}) \ldots (a_{R_{s-2}} - \bar{a})((1-\bar{a}) P(a_{R_{s-1}} = 1|T_{N,R_{s-1}}) - \bar{a} P(a_{R_{s-1}} = 0|T_{N,R_{s-1}}))} = \tag{2.204}$$

$$\overline{(a_0 - \bar{a}) \ldots (a_{R_{s-2}} - \bar{a}) \sum_{r=1}^{N} F(r)(a_{R_{s-1}-r} - \bar{a})}.$$

Here notation $T_{N,R_{s-1}}$ is the set of symbols $(a_{R_{s-1}-N}, \ldots, a_{R_{s-1}-1})$. In that way, we obtain the fundamental recurrence relation connecting the correlation functions of different orders s,

$$K_s(r_1, \ldots, r_{s-1}) = \sum_{r=1}^{N} F(r) K_s(r_1, \ldots, r_{s-1} - r). \tag{2.205}$$

In particular case $s = 2$, this equation turns into Eq. (2.109) that was obtained and discussed in Subsection 2.2.1. Recurrence relation (2.205) is correct for $r_i > 0, i = 1, \ldots, s-1$.

In the case $r_{s-1} \leqslant N$, the last argument of the correlation function in the right hand side of Eq. (2.205) is negative or zero and we should interpret it in the following way, which is referred to as *"collating"*. If the correlation function has negative arguments, we must reorganize it according to definition (1.5). For example,

$$K_4(2, 2, -3) = \overline{(a_0 - \bar{a})(a_2 - \bar{a})(a_4 - \bar{a})(a_1 - \bar{a})} =$$

$$= \overline{(a_0 - \bar{a})(a_1 - \bar{a})(a_2 - \bar{a})(a_4 - \bar{a})} = K_4(1, 1, 2). \tag{2.206}$$

If the correlation function has zero arguments (indexes i and k of two multipliers $(a_i - \bar{a})$ and $(a_k - \bar{a})$ coincide) we should use the property of the *binary* chain: $a_i^2 = a_i$, $a_i = \{0, 1\}$. Thus, we have

$$(a_i - \bar{a})^2 = (1 - 2\bar{a})(a_i - \bar{a}) + \bar{a}(1 - \bar{a}).$$

Using this property we can write very useful relations for the correlation function containing zero arguments in different positions among all arguments of K_s,

$$\begin{aligned} K_s(0, r_2, \ldots, r_{s-1}) &= (1 - 2\bar{a})K_{s-1}(r_2, \ldots, r_{s-1}) + \bar{a}(1 - \bar{a})K_{s-2}(r_3, \ldots, r_{s-1}), \\ K_s(r_1, \ldots, r_{k-1}, 0, r_{k+1}, \ldots, r_{s-1}) &= (1 - 2\bar{a})K_{s-1}(r_1, \ldots, r_{k-1}, r_{k+1}, \ldots, r_{s-1}) \\ &\quad + \bar{a}(1 - \bar{a})K_{s-2}(r_1, \ldots, r_{k-1} + r_{k+1}, \ldots, r_s), \quad k \neq 1, s, \\ K_s(r_1, \ldots, r_{s-2}, 0) &= (1 - 2\bar{a})K_{s-1}(r_1, \ldots, r_{s-2}) + \bar{a}(1 - \bar{a})K_{s-2}(r_1, \ldots, r_{s-3}). \end{aligned} \tag{2.207}$$

The general solution of Eq. (2.205) can be presented as a linear combination,

$$K_s(r_1, \ldots, r_{s-1}) = \sum_{j=1}^{N} L_j(r_1, \ldots, r_{s-2})\xi_j^{r_{s-1}}, \; r_i > 0 \, (i = 1, \ldots, s-1), \tag{2.208}$$

of powers for the roots ξ_j, $j = 1, \ldots, N$, of the characteristic equation,

$$\xi^N - \sum_{j=1}^{N} F(j)\xi^{N-j} = 0. \tag{2.209}$$

New functions $L_j(r_1, \ldots, r_{s-2})$, coefficients of linear form Eq. (2.208), are to be defined in its turn. All one has to do is to substitute Eq. (2.208) into "collated" Eq. (2.205) for $0 < r_{s-1} < N$. Finally, this procedure yields the recurrence relations for the sought functions L_j.

The general solution of this recurrence relation is reduced to the form,

$$L_i(r_1, \ldots, r_{s-2}) = \sum_{k=1}^{N(N-1)/2} M_{ik}(r_1, \ldots, r_{s-3})\eta_k^{r_{s-2}}, \tag{2.210}$$

where η_k, $k = 1, \ldots, N(N-1)/2$, are the roots of the new characteristic equation

$$\det \Upsilon(\eta) = 0, \tag{2.211}$$

$$\Upsilon_{ij}(\eta) = (\eta/\xi_j)^{i-1} - \xi_j^{i-1} + \delta_{1i}, \quad i, j = 1, \ldots, N. \tag{2.212}$$

From the "collating" procedure we can find the next recurrence relations for functions M_{ij} etc. Using this algorithm we can find, in principle, the correlation functions of all orders.

Application of the algorithm. The first simple result is that the correlation functions of all odd orders are zero for the additive Markov chain with $\bar{a}=1/2$. This is resulted from Eq. (2.207): function $K_{2m+1}(\cdot)$ is expressed only via $K_{2m-1}(\cdot)$ and $K_1=0$.

The correlation function of the second order $K_2(\cdot)$ (the *binary* correlation function) for an additive Markov chain can be found from Eq. (2.205) in the form:

$$K_2(r)=\sum_{j=1}^{N} L_j \xi_j^r, \quad r>-N. \tag{2.213}$$

Here ξ_j, $j=1,\ldots,N$, are the roots of the characteristic equation (2.209). Coefficients L_j should be obtained by the collating:

$$K_2(0)=\bar{a}(1-\bar{a}); \quad K_2(-r)=K_2(r), \quad 0<r<N. \tag{2.214}$$

Subsection 2.2.3 contains analysis of these equations in the case of the additive Markov chain with step-wise memory function. Below we present the result of calculation for the correlation function of a weakly correlated chain,

$$|P(a_i=1|a_{i-N}\ldots a_{i-1})-\bar{a}|\ll 1. \tag{2.215}$$

This case is very important from a physical point of view because the statistical properties of the equilibrium long-range correlated Ising chains can be represented and examined as Markov chains. Namely, the Markov chain with a small memory function is statistically equivalent to the weakly correlated Ising chain (see Subsection 2.4).

For the correlation function of even order of the unbiased ($\bar{a}=1/2$) additive chain we have

$$K_2(r)\approx\frac{1}{4}F(r), \quad 0<r<N, \quad K(0)=\frac{1}{4}, \tag{2.216}$$

$$K_{2s}(r_1,\ldots,r_{2s-1})=\prod_{j=1}^{s} K_2(r_{2j-1}), \quad r_j>0,\ j=1,\ldots,2s-1. \tag{2.217}$$

The last equation is correct for the non-additive sequence as well. In the case of one-step Markov chain, $N=1$, this equation is exact. But if $N>1$ we should interpret it as the asymptotical equality. At $N=2$, the *exact* result for the correlation function $K_4(\cdot)$ of the additive chain is,

$$K_4(r_1,r_2,r_3)=K_2(r_1)K_2(r_3)+(-\xi_1\xi_2)^{r_2}\left(\frac{1}{4}K_2(r_1+r_3)-K_2(r_1)K_2(r_3)\right). \tag{2.218}$$

Here ξ_1 and ξ_2 are the roots of the characteristic equation,

$$\xi^2-F(1)\xi-F(2)=0. \tag{2.219}$$

At last, Eq. (2.205) allows us to express the memory function via the prescribed correlation functions K_s. The method of constructing the correlated binary sequence with prescribed binary correlation function K_2 will be discussed in Subsection 3.2.

Generalized algorithm of finding the correlation functions. Here we generalize the algorithm for the additive Markov chains to the arbitrary binary multi-step Markov chains.

Let us write the conditional probability function of the N-step binary Markov chain in the form:

$$P(a_i = 1 | a_{i-N} \dots a_{i-1}) = \sum_{\substack{l_j=0,1 \\ j=1,\dots,N}} F(l_1, l_2, \dots, l_N) \prod_{r=1}^{N} (a_{i-r} - \bar{a})^{l_r}. \tag{2.220}$$

It is a general form for the arbitrary binary function, because $P(\cdot|\cdot)$ can be considered as a linear function of each argument a_j. We refer to function $F(l_1, l_2, \dots, l_N)$ as the *generalized* memory function.

Equation (2.220) can be applied to the additive Markov chain described by Eq. (1.2), namely,

$$\begin{gathered} F(0,0,\dots,0) = \bar{a}, \\ F(\underbrace{0,\dots,0}_{r-1},1,\underbrace{0,\dots,0}_{N-r}) = F(r), \quad r = 1,2,\dots,N, \\ F(l_1,l_2,\dots,l_N) = 0, \quad l_1 + l_2 + \dots + l_N > 1. \end{gathered} \tag{2.221}$$

Now we obtain the recurrence relation for the correlation function with the coefficients expressed via the generalized memory function. As in the beginning of this Subsection, we substitute the last symbol $a_{r_1+\dots+r_{s-1}}$ in Eq. (1.5) (for $r_j > 0$, $j = 1, \dots, s-1$) for its conditional probability $P(\cdot|\cdot)$ in the form of Eq. (2.220). As a result we get,

$$\begin{aligned} K_s(r_1, \dots, r_{s-1}) = (F(0,0,\dots,0) - \bar{a}) K_{s-1}(r_1, \dots, r_{s-2}) + \\ + \sum_{\substack{l_j=0,1 \\ j=1,\dots,N}} F(l_1, l_2, \dots, l_N) K_{s+k-1}(r_1, \dots, r_{s-2}, r_{s-1} - \rho_k, \rho_k - \rho_{k-1}, \dots, \rho_2 - \rho_1). \end{aligned} \tag{2.222}$$

Here $k = l_1 + \dots + l_N \neq 0$. The increasing numbers ρ_j, $j = 1, \dots, k$ are the indexes, for which $l_{\rho_j} \neq 0$.

Note that some summands in Eq. (2.222) can have non-positive arguments in the case of $r_{s-1} \leqslant N$. For this case, we should apply the "collating" procedure to this equation.

Thus, the algorithm of finding of the correlation functions can be formulated in following form:

1. To obtain the correlation function $K_s(r_1, r_2, \dots, r_{s-1})$ at $r_i > 0$, $i = 1, 2, \dots, s-1$, we should use Eq. (2.222) and execute the "collating" procedure. We find that function $K_s(r_1, \dots, r_{s-1})$ is expressed via the correlation functions with the sum of their arguments less than initial sum $r_1 + \dots + r_{s-1}$.
2. Using Eq. (2.222) and performing the "collating" procedure m times with respect to obtained correlation functions, we derive a relation between function $K_s(r_1, \dots, r_{s-1})$ and some other correlation functions.
   a) If all of them have the order less than s, we obtain the recurrence relation for the correlation function of the sth order. This relation can be solved by the method of characteristic equations without executing item 3. (A similar case takes place for the additive Markov chain.)

b) In the opposite case, for $m > r_1 + r_2 + \cdots + r_{s-1} - N$, function $K_s(r_1, \dots, r_{s-1})$ is expressed via the correlation functions with the sums of their arguments less than N. Some of these functions are of the order greater than s. Then we should go to the next item of the algorithm.

3. All correlation functions with the sum of arguments less than N should be found from a set of linear equations. For this purpose we write Eq. (2.222) for every such correlation function. After executing the "collating" procedure we obtain the set of $(2^{N-1} - 1)$ linear equations for $(2^{N-1} - 1)$ sought values of the correlation functions.

In some particular cases, it is convenient to define $K_s(\cdot)$ in Eq. (1.5) and $P(\cdot|\cdot)$ in Eq. (2.220) using an arbitrary fixed value $\tilde{a}$ (for example, $\tilde{a} = 1/2$) instead of the average $\bar{a}$. In this case, all reasonings remain valid provided that the value of $\bar{a}$ is changed to $\tilde{a}$ in Eqs. (2.222) and (2.207), and the condition $K_1 = 0$ is changed to $K_1 = \tilde{a} - \bar{a}$. For obtaining average $\bar{a}$ we should add Eq. (2.222), written for K_1, to the set of equations in item 3 of the algorithm. Then we obtain the set of 2^{N-1} linear equations and $(2^{N-1} - 1)$ sought values of the correlation functions and one of the average $\bar{a}$. Besides, we can use value $\tilde{a}$ instead of average $\bar{a}$ in the different physical problems, which are concerned with the Markov chain. For instance, the energy in the Ising model is more convenient to express via function $K_2(r) = \overline{(a_i - 1/2)(a_{i+r} - 1/2)}$.

2.3.4 Chains with a Nonlinear Memory Function

The method developed in the previous Subsection can be applied to the problem of constructing a Markov chain with the same correlation functions as a given sequence. If we are interested in the binary correlation function only the sequence can be simulated by the additive Markov chain. Its memory function can be found from Eq. (2.109) as mentioned in Subsection 2.2.1 and will be discussed in more detail in Subsection 3.2.

Yet another approach to numerical finding the memory function is an iteration procedure that is especially useful if the correlations in the chain are weak. We start from reconstructing the chain with given binary correlation function. For the realization of iteration procedure let us rewrite Eq. (2.109) in the form,

$$F(r) = \frac{K(r)}{K(0)} - \sum_{r'=1,\ r'\neq r}^{N} \frac{K(r-r')}{K(0)} F(r'). \tag{2.223}$$

Using Eq. (2.223) with starting iteration $F_0(r) = 0$, we obtain the formula,

$$F_{n+1}(r) = \frac{K(r)}{K(0)} - \sum_{r'=1,\ r'\neq r}^{N} \frac{K(r-r')}{K(0)} F_n(r'),\ n \geqslant 0. \tag{2.224}$$

Thus, the memory function can be presented as the series,

$$\begin{aligned} F(r) = {} & \frac{K(r)}{K(0)} - \sum_{r'\neq r} \frac{K(r-r')K(r')}{K^2(0)} \\ & + \sum_{r'\neq r} \sum_{r''\neq r'} \frac{K(r-r')K(r'-r'')K(r'')}{K^3(0)} + \dots \end{aligned} \tag{2.225}$$

Note, that the Markov chain with the definite correlation function $K(r)$ exists if the series (2.225) is convergent and the obtained function implies the probability Eq. (1.2) satisfying the requirement $0 \leqslant P(a_i = 1 \mid T_{N,i}) \leqslant 1$ for arbitrary word $T_{N,i}$. If $\bar{a} = 1/2$ we obtain the restriction, $\sum |F(r)| \leqslant 1$. The sufficient, but not necessary, requirement is $\sum_{r=1}^{N-1} |K(r)| \leqslant 1/12 - |K(N)|/3$.

Now let us construct the Markov chain possessing the same correlation functions of the second and third orders as a given system, [66]. We assume that the normalized binary correlation function

$$\mathcal{K}_2(j) = (\overline{a_i a_{i+j}} - \bar{a}^2)/\bar{a}(1-\bar{a})$$

and the normalized correlation function of the third order

$$\mathcal{K}_3(j,l) = \overline{(a_i - \bar{a})(a_{i-l} - \bar{a})(a_{i-j-l} - \bar{a})}/\bar{a}(1-\bar{a})$$

are much less than unity, i.e. inequalities

$$\sum_j |\mathcal{K}_2(j)| \ll 1, \quad \sum_{j,l} |\mathcal{K}_3(j,l)| \ll 1$$

are fulfilled. We seek the Markov chain with the conditional probability of the form

$$P(a_i = 1 | a_{i-N}, \ldots, a_{i-1}) = F_0 + \sum_{j>0} F_1(j)(a_{i-j} - \bar{a})$$
$$+ \sum_{m>l>0} F_2(l,m)(a_{i-l} - \bar{a})(a_{i-m} - \bar{a}), \tag{2.226}$$

that takes into account the three-element non-linear interaction between symbols. It is evident that this form is the special case of Eq. (2.220), yet it is more general than the case of additive chain with conditional probability Eq. (1.2). We use the general algorithm of high order correlation functions calculation as well as the perturbation theory. The conditional probability function Eq. (2.226) was found asymptotically in the small parameter $\mathcal{K}_2(j)$:

$$F_0 = \bar{a}, \quad F_1(j) = \mathcal{K}_2(j) - \sum_{l<j} \mathcal{K}_2(j-l)\mathcal{K}_2(l),$$
$$F_2(j,l) = \frac{1-2\bar{a}}{\bar{a}(1-\bar{a})}(\mathcal{K}_3(l-j,j) - \mathcal{K}_2(l-j)(\mathcal{K}_2(j) + \mathcal{K}_2(l))). \tag{2.227}$$

Thus we showed the method of preparing the system with weak but long-range interaction between its elements with the prescribed correlation functions of the second and third orders.

We have presented the method to find different statistical properties of the N-step Markov chain. It is characterized by the probability for the symbol to take on some definite value under the condition of the fixed value of the *previous* symbols. Though such a way of determining of a stochastic system is natural if index i of the symbol a_i represents the quantized time, for example, if we *generate* a sequence with prescribed statistical properties. But in different applications, especially in physical, the index i play role of the spatial coordinate or time. In these cases it is easier to find the probability under the condition of *all* the fixed symbols in the chain. In the next Subsection we investigate such systems and show their equivalence with studied Markov chains, [67].

2.4 EQUIVALENCE OF MARKOV AND TWO-SIDED CHAINS

In different mathematical and physical problems, we confront the random sequences for which the probability of symbol a_i to have a certain value under the condition that the values of the **rest** of symbols are fixed depends on the values of not only N **previous** symbols but on the values of N **next** symbols as well,

$$P(a_i = a| \dots , a_{i-2}, a_{i-1}, a_{i+1}, a_{i+2}, \dots) \\ = P_{2,N}(a_i = a|a_{i-N}, \dots , a_{i-1}, a_{i+1}, \dots , a_{i+N}). \tag{2.228}$$

We call such chains N-*two-sided* chains. We can define *additive* binary two-sided chains similarly to the additive Markov ones,

$$P_{2,N}(a_i = 1|a_{i-N}, \dots , a_{i-1}, a_{i+1}, \dots , a_{i+N}) = \bar{a} + \sum_{\substack{r=-N \\ r\neq 0}}^{N} G(r)(a_{i+r} - \bar{a}). \tag{2.229}$$

Here $G(r)$, $r = \pm 1, \dots , \pm N$, is the *memory function of the two-sided chain*. This function characterizes not only the memory, but also the *forecast* function. However in the sake of simplicity we will omit the term forecast.

For the same reason as for the Markov chains, the two-sided sequences are homogeneous. We also suppose that this chain has the property of the metric transitivity, i.e.,

$$0 < P_{2,N}(\cdot|\cdot) < 1, \tag{2.230}$$

for all values of its arguments.

There are no simple methods for constructing the two-sided chains. The best known and simple approach is the Metropolis algorithm [68]. However, it requires many more operations than for constructing the Markov chains. Therefore, it is additional reason why it is very important to prove the equivalence of the Markov and two-sided chains.

In this Subsection, we prove the equivalence of two random sequences, the Markov and two-sided chains. The proof is given for a binary chain, $a_i = \{0, 1\}$, but it can be directly generalized to arbitrary chains. We do this in Subsection 2.4.3. The proof requires some formulas for the conditional probability. Its definition is

$$P(A|B) = \frac{P(A, B)}{P(B)}. \tag{2.231}$$

Here and subsequently, a comma between the symbols-events implies that both of these events occur simultaneously, i.e., it is a product of two events, $(A, B) = A \cap B$. Using evident equation,

$$P(A, B|C) = P(A|B, C)P(B|C), \tag{2.232}$$

the following formula can be easily derived:

$$P(A|B, C) = \frac{P(A, B|C)}{P(A, B|C) + P(\overline{A}, B|C)}, \tag{2.233}$$

where $\overline{A}$ is the event complementary to A.

2.4.1 Transition from the Markov to two-sided Chains

Let us demonstrate that any binary Markov chain is a two-sided chain. For this purpose, using Eq. (2.233), we rewrite the probability that symbol a_i is equal to unity, under the condition that the **rest** of symbols in the Markov chain are given, in the form,

$$P(a_i=1|A_i^-, A_i^+) = \frac{P(a_i=1, A_i^+|A_i^-)}{P(a_i=1, A_i^+|A_i^-)+P(a_i=0, A_i^+|A_i^-)}. \tag{2.234}$$

Here A_i^- and A_i^+ are the semi-infinite words, $(\dots, a_{i-2}, a_{i-1})$ and $(a_{i+1}, a_{i+2}, \dots)$, respectively,

$$\underbrace{\dots, a_{i-2}, a_{i-1}}_{A_i^-}, a_i, \underbrace{a_{i+1}, a_{i+2}, \dots}_{A_i^+}. \tag{2.235}$$

The main idea of our proof is to express probabilities $P(a_i, A_i^+|A_i^-)$ containing semi-infinite words in terms of the conditional probability function of the Markov chain. For this, we need to use iteratively Eq. (2.232) in order to present $P(\cdot|\cdot)$ in factor form,

$$P(a_i=1, A_i^+|A_i^-) = P(a_i=1|A_i^-)P(a_{i+1}, A_{i+1}^+|a_i=1, A_i^-) =$$

$$= P(a_i=1|A_i^-)P(a_{i+1}|a_i=1, A_i^-)P(a_{i+2}, A_{i+2}^+|A_i^-, a_i=1, a_{i+1}) = \dots$$

$$\dots = \prod_{\substack{r=0\\a_i=1}}^{N} P(a_{i+r}|A_{i+r}^-) \prod_{\substack{r=N+1\\a_i=1}}^{\infty} P(a_{i+r}|A_{i+r}^-). \tag{2.236}$$

The sequence under consideration is the N-step Markov chain and, according to definition (1.1), the probability of symbol a_i to be equal to unity depends on the values of N previous symbols only. So, the factors with $r>N$ in the second multiplier $\prod_{r=N+1}^{\infty}$ of Eq. (2.236) are a_i-independent. Substituting expression (2.236) for $P(a_i=1, A_i^+|A_i^-)$ and $P(a_i=0, A_i^+|A_i^-)$ into Eq. (2.234), making the cancelation of an infinite number of identical terms with $r>N$ in the numerator and denominator of Eq. (2.234)[1], we derive the following equation:

$$P(a_i=1|A_i^-, A_i^+) = \frac{\prod\limits_{\substack{r=0\\a_i=1}}^{\infty} P(a_{i+r}|A_{i+r}^-)}{\prod\limits_{\substack{r=0\\a_i=1}}^{\infty} P(a_{i+r}|A_{i+r}^-) + \prod\limits_{\substack{r=0\\a_i=0}}^{\infty} P(a_{i+r}|A_{i+r}^-)}$$

$$= \frac{\prod\limits_{\substack{r=0\\a_i=1}}^{N} P_N(a_{i+r}|T_{i+r}^-)}{\prod\limits_{\substack{r=0\\a_i=1}}^{N} P_N(a_{i+r}|T_{i+r}^-) + \prod\limits_{\substack{r=0\\a_i=0}}^{N} P_N(a_{i+r}|T_{i+r}^-)} = P_{2,N}(a_i=1|T_i^-, T_i^+). \tag{2.237}$$

[1] To be more strict we need to consider a finite chain and then to make a limit transition to the infinite chain.

Here T_i^- and T_i^+ are the previous and next (with respect to symbol a_i) N-words

$$\dots, a_{i-N-1}, \underbrace{a_{i-N} \dots, a_{i-2}, a_{i-1}}_{T_i^-}, a_i, \underbrace{a_{i+1}, a_{i+2}, \dots, a_{i+N}}_{T_i^+}, a_{i+N+1} \dots . \tag{2.238}$$

Equation (2.237) is the fundamental relation for identification of the Markov and two-sided chains. We can see that the probability of symbol a_i to be equal to unity on condition that the **rest** symbols are fixed is determined by two words of the length N, T_i^- and T_i^+, only. So, according to definition Eq. (2.228), the Markov chain is, at the same time, the two-sided, quod erat demonstrandum.

Equation (2.237) makes it possible to calculate easily two-sided conditional probability function $P_{2,N}(a_i|T_i^-, T_i^+)$ via conditional probability function $P_N(a_i|T_i^-)$ of the Markov chain.

2.4.2 From the two-sided to Markov Chain

Here we prove the opposite statement: any two-sided chain is a Markov chain. In other words, we prove that the probability of symbol a_i to be equal to unity, under condition that all **previous** symbols are given, depends on the values of N previous symbols only. For that, we consider two sets of symbols, A' and A'', that are two variants of word A_{i-N}^- and differ only by one symbol a_{i-N-k} at arbitrary $k > 0$,

$$A_{i-N}^-|_{a_{i-N-k}=0} = A', \quad A_{i-N}^-|_{a_{i-N-k}=1} = A''. \tag{2.239}$$

Using the definition of conditional probability, Eq. (2.231), we have

$$P(a_i|A', T_i^-) = \frac{P(A', T_i^-, a_i)}{P(A', T_i^-)} = \frac{P(a'_{i-N-k}|\tilde{A}, T_i^-, a_i)P(\tilde{A}, T_i^-, a_i)}{P(a'_{i-N-k}|\tilde{A}, T_i^-)P(\tilde{A}, T_i^-)} =$$

$$= \frac{P(a'_{i-N-k}|\tilde{A}, T_i^-, a_i)}{P(a'_{i-N-k}|\tilde{A}, T_i^-)} P(a_i|\tilde{A}, T_i^-), \tag{2.240}$$

where $\tilde{A}$ is the set of symbols of word A_{i-N}^- without symbol a_{i-N-k}, "overbraced" in Eq. (2.241),

$$\underbrace{\overbrace{\dots, a_{i-N-k-1}}^{\tilde{A}}, a_{i-N-k}, \overbrace{a_{i-N-k+1} \dots, a_{i-N-1}}^{\tilde{A}}}_{A_{i-N}^-}, \underbrace{a_{i-N} \dots, a_{i-2}, a_{i-1}}_{T_i^-}, a_i, \dots . \tag{2.241}$$

Conditional probability $P(a'_{i-N-k}|\tilde{A}, T_i^-, a_i)$ does not depend on symbol a_i, according to definition Eq. (2.228) of two-sided chain, since the distance between a_i and a_{i-N-k} is greater than N. Hence we get

$$P(a_i|A', T_i^-) = P(a_i|\tilde{A}, T_i^-) = P(a_i|A'', T_i^-). \tag{2.242}$$

With the help of this result we can prove that conditional probability $P(a_i|A_{i-N}^-, T_i^-)$ is independent of the values of two, three and so on symbols in word A_{i-N}^-. Thus, we conclude that probability $P(a_i|A_i^-)$ takes on the same value for any arbitrary word A_{i-N}^-, i.e., it does not depend on A_{i-N}^-. Thus, we make sure that

$$P(a_i|A_i^-) = P_N(a_i|T_i^-). \tag{2.243}$$

In other words, according to definition (1.1), the two-sided chain is a Markov chain, quod erat demonstrandum.

It should be noticed that each two-sided sequence is equivalent to the **unique** Markov chain. This statement is not evident from Eq. (2.237) because of the non-linear structure of this equation. Using trivial variant of Eq. (2.231),

$$P_N(a_i|T_i^-) = \frac{P(T_i^-, a_i)}{P(T_i^-)}, \tag{2.244}$$

we can easily see that a given chain cannot have two different conditional probabilities. The point is that the probabilities of N- and $(N+1)$-words occurring determine the conditional probability function unambiguously according to Eq. (2.244). Hence, for the chain under study, the Markov conditional probability is also determined uniquely.

The non-linear structure of Eq. (2.237) presents some difficulties to calculate conditional probability function $P_N(a_i|T_i^-)$ of the Markov chain by means of two-sided conditional probability function $P_{2,N}(a_i|T_i^-, T_i^+)$.

2.4.3 Non-Binary Chain

The results obtained in Subsections 2.4.1 and 2.4.2 can be generalized to non-binary chains. The following equation connecting the conditional probability functions can be derived:

$$P_{2,N}(a_i = a|T_i^-, T_i^+) = \frac{\prod\limits_{\substack{r=0 \\ a_i=a}}^{N} P_N(a_{i+r}|T_{i+r}^-)}{\sum\limits_{\xi \in \mathcal{A}} \prod\limits_{\substack{r=0 \\ a_i=\xi}}^{N} P_N(a_{i+r}|T_{i+r}^-)}, \tag{2.245}$$

which is analogue of Eq. (2.237). If symbol a takes on the continuous set of values, $\mathcal{A}$, summation $\sum\limits_{\xi \in \mathcal{A}}$ in this formula should be replaced by integration $\int\limits_{\mathcal{A}} d\xi$.

Thus, the equivalence between the N-two-sided and N-step Markov chains takes place for non-binary chains as well. We found the important formula for the conversion of the Markov's conditional probability to the two-sided one, and vice versa. This method can be used for numerical and analytical calculations of the conditional probabilities. Let us note that it is easier calculate $P_{2,N}(a_i = a|T_i^-, T_i^+)$ knowing the conditional probability function $P_N(a_{i+r}|T_{i+r}^-)$, than to solve the inverse problem. In the last case we need to solve the set of nonlinear equations.

We have proved that the classes of the usual "one-sided" multi-step Markov chains and two-sided sequences coincide. The obtained relationship between the conditional probability functions allows us to construct numerically the Markov chain possessing the same statistical properties as the initial two-sided one. Any Markov chain can be characterized not only by the conditional probability $P_N(a_i = a|T_i^-)$, but by the two-sided conditional probability function, $P_{2,N}(a_i = a|T_i^-, T_i^+)$, as well.

2.5 ISOTROPY PROPERTIES

As mentioned in the previous Subsection the established correspondence between the Markov and two-sided chains allows us to study some new statistical characteristics of the sequences with spatial correlations. In this Subsection, we examine isotropy property of the Markov chains. Let us define an *isotropic* chain as a sequence for which the probability of an arbitrary set of L sequent symbols occurring, referred to as the *L-word*, does not depend on the direction of "reading" a chain. For example, if we take 5-word 01100, then its probability occurring equals the probability to have the word 00110.

2.5.1 Two-Sided Chains with Symmetric Probability Function

We start our study from the important class of Markov chains that are equivalent to the two-sided sequences with symmetric conditional probability function. The given definition of isotropic N-step Markov chain is equivalent to the following statement: the Markov chain is isotropic if its two-sided conditional probability function is symmetrical,

$$P_{2,N}(a_i = a|T^-_{\rightarrow}, T^+_{\rightarrow}) = P_{2,N}(a_i = a|T^+_{\leftarrow}, T^-_{\leftarrow}). \tag{2.246}$$

Here T^- and T^+ are previous and next with respect to symbol a N-words. The subscript "$_\rightarrow$" indicates that word T is read in the direct order, from the left to the right, and the subscript "$_\leftarrow$" shows that word T is read in the inverse order. For example, if the binary 3-two-sided chain is isotropic then

$$P_{2,3}(a_i = 1|011, 001) = P_{2,3}(a_i = 1|100, 110).$$

Equation (2.246) can serve as the second definition of the isotropy for the Markov chains.

Below we prove the equivalence of these two definitions of the isotropy.

Suppose that the first definition is valid and, therefore, the probability $P(T^-_{\rightarrow}, a_j = a, T^+_{\rightarrow})$ of an arbitrary $(2N+1)$-word occurring in the Markov chain is independent of the direction of its reading. Then, the formula for two-sided probabilities can be rewritten as

$$\begin{aligned} P_{2,N}(a_j = a|T^-_{\rightarrow}, T^+_{\rightarrow}) &= \frac{P(T^-_{\rightarrow}, a_j = a, T^+_{\rightarrow})}{\sum\limits_b P(T^-_{\rightarrow}, a_j = b, T^+_{\rightarrow})} \\ &= \frac{P(T^+_{\leftarrow}, a_j = a, T^-_{\leftarrow})}{\sum\limits_b P(T^+_{\leftarrow}, a_j = b, T^-_{\leftarrow})} = P_{2,N}(a_j = a|T^+_{\leftarrow}, T^-_{\leftarrow}). \end{aligned}$$

Thus, the two-sided conditional probability function of the Markov chain is symmetrical and the chain is isotropic in accordance with the second definition.

Now, let us suppose that the second definition of the isotropy is valid, i.e., the two-sided probability of the Markov chain is symmetrical.

So, this chain, being read in direct and inverse orders, is statistically the same. In other words, the probability of symbol $a_i = a$ occurring under condition that **previous** N-word $T_{\rightarrow}$ is fixed equals the probability of the same symbol occurring under condition that **next** N-word $T_{\leftarrow}$ is fixed. It means that the original chain and its copy written in inverse order have the same Markov conditional probabilities. However, the probabilities of words of arbitrary length L are determined completely by the conditional probabilities. Hence, these probabilities are the same for both chains.

Thus, the probability of an arbitrary word of length L occurring does not depend on the direction of its reading.

According to Eq. (2.237) this chain, being read in direct and inverse orders, has the same Markov conditional probabilities. However, the probabilities of words of arbitrary length L are determined completely by the conditional probabilities. Hence, this chain is isotropic according the first definition.

In the general case, the Markov chains are anisotropic. Nevertheless, a consideration of Eq. (2.237) shows that all 1-step and 2-step Markov chains are isotropic.

For instance, let us show that the conditional probabilities $P(a_i = 1|11, 01)$ and $P(a_i = 1|10, 11)$ are equal to each other. Indeed, Eq. (2.237) yields for these probabilities the following expressions:

$$P(a_i = 1|11, 01) = \left(1 + \frac{(1 - P(1|11))(1 - P(1|10))P(1|00)}{P(1|11)(1 - P(1|11))P(1|10)}\right)^{-1}, \tag{2.247}$$

$$P(a_i = 1|10, 11) = \left(1 + \frac{(1 - P(1|10))P(1|00)P(1|01)}{P(1|10)P(1|01)P(1|11)}\right)^{-1}. \tag{2.248}$$

The analogous calculations can be carried out for all other combinations of arguments in the two-sided conditional probability function. So all 2-step Markov chains (and, hence, all 1-step Markov chains) are isotropic.

All **non-biased** 3-step additive Markov chains are isotropic also. The additive chains can be anisotropic for $N \geqslant 3$. The 3-step biased Markov chains with conditional probability

$$P(a_i = 1|a_{i-3}, a_{i-2}, a_{i-1}) = \bar{a} + \sum_{r=1}^{3} f_r(a_{i-r} - \bar{a}) \tag{2.249}$$

and $\bar{a} \neq 1/2$ are isotropic only in three exceptional cases:

1. $f_1 = f_2$. This condition is fulfilled, e.g., for the chains with the step-wise memory function.
2. $f_1 + f_2 + f_3 = 1$. It is the degenerated case. Such a memory function determines the Markov chain consisting completely of unities or zeroes, since $P(a_i = 1|111) = P(a_i = 0|000) = 1$.
3. $f_3 = 0$. Actually, it is the 2-step additive Markov chain.

All additive Markov chains with small memory functions, $F(r) \propto \varepsilon \ll 1$, are isotropic in the main approximation with respect to ε. This fact will be clarified in Subsection 2.5.3.

It should be noted that the two-sided chain with asymmetrical conditional probability function can be considered as Markov chain being read in the direct order and as another Markov chain being read in the inverse order. The conditional probability functions of these chains are different. Two conclusions follow from this consideration:

1. There are at least two different asymmetrical N-steps Markov chains with equal correlation functions,

$$K(r) = \overline{(a_i - \bar{a})(a_{i+r} - \bar{a})}. \tag{2.250}$$

2. The additive anisotropic Markov chain being read in the inverse order is a non-additive Markov chain. Otherwise we would have two additive Markov chains with different memory functions but with the same correlation function. As it follows from Subsection 2.2.3 it is impossible.

2.5.2 Permutative Markov Chains

Now we investigate the second class of Markov chains, permutative chains, that were introduced in Subsection 2.3.2.

Isotropy of permutative Markov chains. Here we prove that the permutative binary N-step Markov chain is isotropic. Let us consider two chains, M and M$'$, where M is the given Markov chain, $\{a_i\}$, and M$'$ is the chain written in the inverse order, $\{a'_i\}$, $a'_i = a_{-i}$. We refer to $\overrightarrow{P}_k$, or $\overrightarrow{P}(a_i = 1|T(k)) = \overrightarrow{P}(1|T(k))$, as the probability of symbol a_i in chain M to be equal unity under condition that the previous N-word $T(k)$ contains k unities. The probability of the N-word occurring satisfies the set of linear equations,

$$\begin{cases} P(a_1, a_2, \dots, a_N) = \sum\limits_{a_0} \overrightarrow{P}(a_N|a_0, a_1, \dots, a_{N-1}) P(a_0, a_1, \dots, a_{N-1}), \\ \sum\limits_{a_1} \dots \sum\limits_{a_N} P(a_1, a_2, \dots, a_N) = 1. \end{cases} \tag{2.251}$$

The probability $P(a_1, a_2, \dots, a_N)$ of the N-word occurring is determined uniquely by Eq. (2.251). The solution of this set of equations depends on the total number $k = a_1 + a_2 + \dots + a_N$ of the unities in the N-word $(a_1, a_2, \dots, a_N)$ and can be presented in the form $P(a_1, a_2, \dots, a_N) = P_k$. We can easily find the following expression for P_k,

$$P_k = P_0 \prod_{r=1}^{k} \frac{\overrightarrow{P}_{r-1}}{1 - \overrightarrow{P}_r}.$$

We also refer to $\overleftarrow{P}(1|T)$ as the probability of symbol a'_i to be equal unity under condition that the previous N-word in chain M$'$, T, is fixed. Now let us prove that the conditional probability functions of chains M and M$'$ are equal to each other. In other words, $\overleftarrow{P}(1|T)$ equals to $\overrightarrow{P}_k$ for arbitrary N-word T containing exactly k unities ($k = 1, \dots, N$) and does not depend on the order of symbols in this word. With this aim we need to prove that

$$\overleftarrow{P}(1|[k], 0) = \overleftarrow{P}(1|[k-1], 1) = \overrightarrow{P}_k.$$

Here $[j]$ means $(N-1)$-word containing exactly j unities, so that $(1, [k-1])$ and $([k-1], 1)$ are the N-words containing k unities. Using the definition of the conditional probability we find:

$$\overleftarrow{P}(1|[k-1], 1) = \overleftarrow{P}(1|[k-1], 1)\frac{P_k}{P_k} = \frac{P(1, [k-1], 1)}{P_k} = \overrightarrow{P}(1|1, [k-1]) = \overrightarrow{P}_k,$$

$$\overleftarrow{P}(1|[k], 0) = \frac{P(1, [k], 0)}{P_k} = (1 - \overrightarrow{P}_{k+1})\frac{P_{k+1}}{P_k}.$$

From set (2.251), we get the relation

$$P_{k+1} = \overrightarrow{P}_{k+1} P_{k+1} + \overrightarrow{P}_k P_k.$$

So, we proved that

$$\overleftarrow{P}(1|[k-1], 1) = \overleftarrow{P}(1|[k], 0) = \overrightarrow{P}_k.$$

Thus we can refer to $\overleftarrow{P}_k$ as the probability of the symbol in chain M′ to be equal unity on condition that the previous N-word containing exactly k unities is fixed, and the conditional probability functions for the chains M and M′ are equal to each other,

$$\overrightarrow{P}_k = \overleftarrow{P}_k, \; k \geqslant 1. \tag{2.252}$$

Hence,

$$\overrightarrow{P}_0 = \overleftarrow{P}_0 \tag{2.253}$$

since

$$\sum_T \overrightarrow{P}(1|T)P(T) = \sum_T \overleftarrow{P}(1|T)P(T) = 1.$$

Equations (2.252) and (2.253) imply actually that the permutative binary N-step Markov chains are isotropic. The special case of this statement was a corollary from a Proposition 1 in Subsection 2.1.2 in the case of additive chains with step-wise memory function.

Two-sided probability functions of permutative Markov chains. As mentioned above (see Eq. (2.237)), every Markov chain can be considered as two-sided chain. Below we study the properties of the two-sided conditional probability function of the Markov chains with one-sided conditional probability functions possessing the property of permutability. The very important fact is that this property does not provide the permutability of the two-sided conditional probability function. To demonstrate this fact we will show that in the general case the two-sided conditional probability function changes its value when two neighboring symbols 1 and 0 are transposed.

Let us prove this statement by contradiction and suppose that the two-sided conditional probability function $P(a_i = 1|T_i^-, T_i^+)$ takes on the same value for two variants of the word T_i^-. These variants of the word T_i^- differ only by the values of the neighboring symbols a_j and a_{j+1}: $a_j = 1, a_{j+1} = 0$ in the first variant and $a_j = 0, a_{j+1} = 1$ in the second variant. Consider the structure of Eq. (2.237). Taking into account the permutability of Markov conditional probability function we can see, that N factors in all products in Eq. (2.237) coincide for two symbols sets under consideration. In the general case, only one factor, $P_N(a_{j+N+1}|T_{j+N+1}^-)$, changes its value. If $a_{j+N+1} = 1$, the coincidence of the values of $P(a_i = 1|T_i^-, T_i^+)$ for two considering sets of symbols yields the following relation between the values of one-sided conditional probability function:

$$\overrightarrow{P}_k^2 = \overrightarrow{P}_{k+1} \overrightarrow{P}_{k-1} \tag{2.254}$$

for $k = a_{j+2} + a_{j+3} \ldots + a_{j+N}$, $1 \leqslant k \leqslant N-1$. In the opposite case, at $a_{j+N+1} = 0$, the similar requirement is

$$(1 - \overrightarrow{P}_k)^2 = (1 - \overrightarrow{P}_{k+1})(1 - \overrightarrow{P}_{k-1}). \tag{2.255}$$

These two relations are compatible for non-correlated chain only where $\overrightarrow{P}_k$ is independent of k. The exception is the case of the one-step binary Markov chain, that always has a permutative two-sided conditional probability function.

Thus, any correlated multi-step permutative Markov chain possesses the non-permutative two-sided probability function.

2.5.3 Additive Weakly Correlated Markov Chains

The third class of the isotropic sequences represents the additive weakly correlated Markov chains. For these chains, we assume that the memory function is small,

$$\sum_{r=1}^{N} |F(r)| \ll 1. \tag{2.256}$$

Their asymptotical isotropy is proved in the first paragraph.

Every function of N variables, $f(a_1,\dots,a_N)$, satisfying the evident restriction, $0 \leqslant f(a_1,\dots,a_N) \leqslant 1$, can be considered as the conditional probability function $P(a_i = 1|a_{i-N},\dots,a_{i-2},a_{i-1})$ of some Markov chain. However, not every function of $2N$ variables, even restricted by the similar condition, can be the conditional probability function of some two-sided chain. As shown in Eq. (2.237), an arbitrary binary Markov chain is determined by 2^N parameters, the number of all possible sets of arguments in the conditional probability function. Hence, a two-sided chain, equivalent to this Markov chain, is also determined by 2^N parameters. Nevertheless, the two-sided conditional probability function $P(a_i = 1|T_i^-, T_i^+)$ is formally determined by 2^{2N} parameters, number of all possible **previous**, T_i^-, and **next**, T_i^+, N-words. So, not every function of $2N$ arguments can play a role in some two-sided conditional probability function. The example is an additive two-sided chain with small memory function $G(r)$ in Eq. (2.229). In this case, $G(r)$ has to be asymptotically even. This fact is proved in the second paragraph.

Isotropy of additive weakly correlated Markov chains. In order to find the conditional probability function, $P(a_i = 1|T_i^-, T_i^+)$, of the weakly correlated additive Markov chain we have to substitute the probability Eq. (1.2) into Eq. (2.237), and retain terms of the zeroth and first orders in $F(r)$. The obtained two-sided conditional probability function takes the form of Eq. (2.229) with even memory function: $G(r) = G(-r) = F(r)$. So, the additive weakly correlated Markov chains are asymptotically isotropic.

Restriction on the class of the memory functions of weakly correlated additive two-sided chains. In this Subsection, we show that the weakly correlated additive two-sided chain is asymptotically isotropic, i.e. the two-sided memory function $G(r)$ is necessarily even. To that end we consider an arbitrary two-sided additive chain and find the corresponding one-sided conditional probability function, $P(a_i = 1|T_i^-)$. We will prove that this function is reduced to the additive form, Eq. (1.2), with the memory function $F(r) = G(-r)$, and, therefore, the chain under consideration is asymptotically isotropic.

Let us consider the additive two-sided chain (not obligatory isotropic) and find the corresponding Markov chain. In the general case the problem is reduced to solving set of 2^N non-linear equations, Eq. (2.237), written for different sequences of symbols. We consider a weakly correlated additive two-sided chain subjected to the restriction $\sum_{r=1}^{N}(|G(-r)| +$

$|G(r)|) \ll 1$. Let us present the conditional probability function of the corresponding Markov chain in more convenient form:

$$P(a_i|T_i^-) = 1 - a_i + (2a_i - 1)(\bar{a} + \varphi(T_i^-)), \tag{2.257}$$

where $\varphi(T_i^-)$ is some new function to be determined. The evident equation $P(a_i = 1|T_i^-) = 1 - P(a_i = 0|T_i^-)$ is fulfilled for Eq. (2.257). Note that if $G(r) \to 0$ the probability $P(a_i = 1|T_i^-) \to \bar{a}$ and the new function tends to zero: $\varphi(T_i^-) \to 0$. Now substituting the conditional probability in the form of Eq. (2.257) into Eq. (2.237), expanding it in the series with respect to the small parameter $\varphi(T_i^-)$ and keeping in the series only the terms not higher than the first order in $\varphi(T_i^-)$, we obtain the approximate expression for the two-sided conditional probability:

$$P(a_i = 1|T_i^-, T_i^+) \simeq \bar{a}\left(1 + (1 - \bar{a})\left(\sum_{\substack{r=0\\a_i=1}}^{N} \psi(a_{i+r})\varphi(T_{i+r}^-) - \sum_{\substack{r=0\\a_i=0}}^{N} \psi(a_{i+r})\varphi(T_{i+r}^-)\right)\right). \tag{2.258}$$

Here new function ψ is introduced,

$$\psi(a_i) = \frac{a_i - \bar{a}}{\bar{a}(1 - \bar{a})} = \begin{cases} 1/(\bar{a} - 1), & a_i = 0 \\ 1/\bar{a}, & a_i = 1. \end{cases} \tag{2.259}$$

But this two-sided chain is the given chain with the conditional probability Eq. (2.229), so finally,

$$\sum_{\substack{r=0\\a_i=1}}^{N} \psi(a_{i+r})\varphi(T_{i+r}^-) - \sum_{\substack{r=0\\a_i=0}}^{N} \psi(a_{i+r})\varphi(T_{i+r}^-)$$

$$\simeq \frac{1}{\bar{a}(1 - \bar{a})} \sum_{r=1}^{N} (G(r)(a_{i+r} - \bar{a}) + G(-r)(a_{i-r} - \bar{a})). \tag{2.260}$$

Calculating the difference between Eq. (2.260) written for $a_{i+N} = 1$ and $a_{i+N} = 0$ we get

$$\varphi(T_{i+N}^-)\big|_{a_i=1} - \varphi(T_{i+N}^-)\big|_{a_i=0} \simeq G(N). \tag{2.261}$$

Substituting Eq. (2.261) into Eq. (2.260), we have

$$\sum_{\substack{r=0\\a_i=1}}^{N-1} \psi(a_{i+r})\varphi(T_{i+r}^-) - \sum_{\substack{r=0\\a_i=0}}^{N-1} \psi(a_{i+r})\varphi(T_{i+r}^-)$$

$$\simeq \frac{1}{\bar{a}(1 - \bar{a})}\left(\sum_{r=1}^{N-1} (G(r)(a_{i+r} - \bar{a}) + G(-r)(a_{i-r} - \bar{a})) + G(-N)(a_{i-N} - \bar{a})\right). \tag{2.262}$$

Now we repeat this procedure. We calculate the difference between equations (2.262) written for $a_{i+N-1}=1$ and $a_{i+N-1}=0$ and substitute the obtained result into Eq. (2.262), and so on. At the last step we obtain

$$\varphi(T_i^-) \simeq \sum_{r=1}^{N} G(-r)(a_{i-r}-\bar{a}). \tag{2.263}$$

The conditional probability of Markov chain equivalent to the given two-sided chain is

$$P(a_i=1|T_i^-) \simeq \bar{a} + \sum_{r=1}^{N} G(-r)(a_{i-r}-\bar{a}). \tag{2.264}$$

According to the previous Subsection such Markov chains are isotropic. Thus we have obtained that the additive two-sided weakly correlated chain can only be isotropic. In other words, the memory function of the additive weakly correlated chain can be only even, $G(-r)=G(r)$.

The equivalence of the Markov and two-sided chains proved in Subsection 2.4 allowed us to study the important property of the Markov chains, their anisotropy. The results of this study are shown in the scheme, where $A \xrightarrow{st} B$ means, that the chains from class A, restricted by the statement st, are the members of class B. The most evident fact is that the Markov chains, that possess a symmetric two-sided conditional probability function, are isotropic. Another important class of the isotropic Markov chains are the sequences with the permutative conditional probability functions. An example of such chain is additive Markov chains with step-wise memory function, that were examined in detail in Reference [58, 69]. The additive weakly correlated chains are also isotropic. Such chains play a key role in the non-extensive thermodynamics of Ising chains of classical spins.

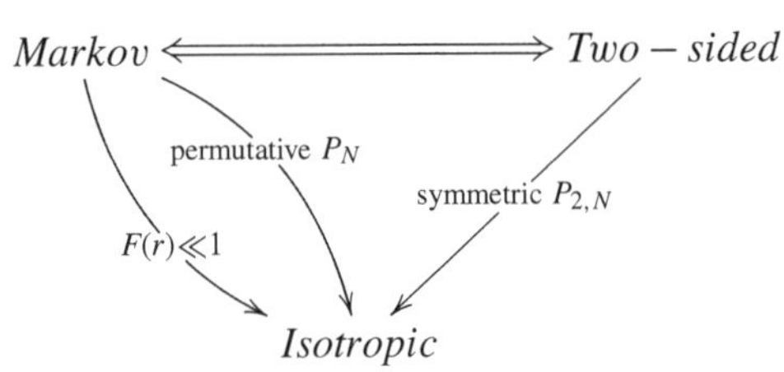

3. Generation of correlated sequences

This Section is devoted to methods of generating random long-range correlated sequences with discrete or continuous space of states. This problem has become more important recently, because much attention was paid to the related problem of how to construct disordered materials with specific transport properties that are due to underlying correlations in a disorder. One of the important applications of this problem is the creation of electron nano-devices, optic fibers, rough surfaces, acoustic and electromagnetic waveguides with selective transport properties. In many applications it is more convenient to construct the sequences with finite number of random values. An important example of such a system, which occurs in nature, is the sequence of nucleotides in a DNA molecule, which consists of four elements only.

First, we describe algorithms to construct a random sequence with the definite designed correlation function, Subsections 3.1–3.4. Furthermore, in Subsection 3.5, some of the known procedures enabling us to construct a sequence with the correlation function of a fixed form, are briefly described.

3.1 CONVOLUTION METHOD

The convolution (Rice) method [19, 71, 70, 22, 25], generating a random sequence with the continuous space of states and desired correlation function, was proposed independently by many authors. Below we describe one of its modifications.

3.1.1 Continuous noise

The convolution method of generation of a *continuous colored noise* $\varepsilon(n)$ starting from a white-noise $\alpha(n)$ is based on a linear transformation with the use of the modulation function $F(n)$, see Refs. [25, 70, 57]. The most general form of this transformation is as follows,

$$\varepsilon(n) = \overline{\varepsilon} + \sqrt{\frac{K_\varepsilon(0)}{K_\alpha(0)}} \sum_{n'=-\infty}^{\infty} F(n-n')[\alpha(n') - \overline{\alpha}]. \tag{3.1}$$

Here $\overline{\alpha}$ and $\overline{\varepsilon}$ are the mean values, and $K_\alpha(0) = \overline{\alpha^2(n)} - (\overline{\alpha})^2$ and $K_\varepsilon(0) = \overline{\varepsilon^2(n)} - (\overline{\varepsilon})^2$ are the variances of the white and colored noise respectively. For a homogeneous sequence $\alpha(n)$ the generated sequence $\varepsilon(n)$ is also homogeneous. In this Subsection, we consider the normalized pair correlator $\mathcal{K}_\varepsilon(r) = K_\varepsilon(r)/K_\varepsilon(0)$, which is an even function of r. Substituting the linear transformation (3.1) into the correlation function $K_\varepsilon(r)$,

$$K_\varepsilon(r) = \overline{\varepsilon(n+r)\varepsilon(n)} - (\overline{\varepsilon})^2 = K_\varepsilon(0)\mathcal{K}_\varepsilon(r) \tag{3.2}$$

and taking into account that the sequence $\alpha(n)$ is δ-correlated, the following relation between the pair correlator and the modulation function is readily obtained,

$$\mathcal{K}_\varepsilon(r) = \sum_{n=-\infty}^{\infty} F(n)F(n+r). \tag{3.3}$$

From the condition $\mathcal{K}_\varepsilon(0) = 1$ we get,

$$\sum_{n=-\infty}^{\infty} F^2(n) = 1. \tag{3.4}$$

Calculating the product $\overline{\varepsilon(n+r)\varepsilon(n)}$, we took into account that the white noise $\alpha(n)$ is ergodic, i.e. the average over n can be replaced by the average over ensemble of white-noise sequences.

3.1.2 Center limit theorem

Since $\mathcal{K}_\varepsilon(n)$ and $F(n)$ are even functions, it is convenient to apply the cosine Fourier transform to both sides of Eq. (3.3). This results in the following relation,

$$\hat{\mathcal{K}}_\varepsilon(k) = \hat{F}^2(k). \tag{3.5}$$

The expression (3.5) determines the modulation function in terms of the Fourier transform of the pair correlator,

$$F(n) = \frac{1}{\pi}\int_0^\pi \hat{\mathcal{K}}_\varepsilon^{1/2}(k)\cos(kn)dk. \tag{3.6}$$

Evidently, the solution (3.6) satisfies the normalization condition (3.4).

For different white-noise sequences $\alpha(n)$ the convolution method, Eqs. (3.1), (2.123) and (3.6), defines an ensemble of colored-noise sequences $\varepsilon(n)$, where $-\infty < \varepsilon(n) < \infty$. All the sequences in this ensemble have the same pair correlator $\mathcal{K}_\varepsilon(n)$. The number of terms contributing to the series (3.1) depends on the sharpness of the correlator $\mathcal{K}_\varepsilon(n)$. For short-range correlations, when $\mathcal{K}_\varepsilon(n)$ decays very fast, the modulation function $F(n)$ is also sharp, therefore, the principal contribution is mainly given by a single term with $n' = n$. The sequence $\varepsilon(n)$ in this case is practically delta-correlated for both continuous and binary sequences $\alpha(n)$.

We are rather interested in analyzing the opposite case of *long-range correlations* when the correlation length R_c is large,

$$R_c \gg 1. \tag{3.7}$$

We shall show that in this case, when the generating function is very smooth and a great number of summands contributes to Eq. (3.1), the distribution function $\rho_E(\varepsilon)$ has the *Gaussian form*. This statement is analogous to the central limit theorem.

A simplest way to display this fact is to calculate *characteristic function* $\varphi_E(t)$ of the random variable $\varepsilon(n)$, which is defined by

$$\varphi_E(t) \equiv \overline{\exp[it\varepsilon(n)]} = \int_{-\infty}^{\infty} d\varepsilon\, \rho_E(\varepsilon) \exp(it\varepsilon). \tag{3.8}$$

From the last equality in definition (3.8), it immediately follows that the desired probability density $\rho_E(\varepsilon)$ is just the Fourier transform of $\varphi_E(t)$,

$$\rho_E(\varepsilon) = \frac{1}{2\pi} \int_{-\infty}^{\infty} dt\, \varphi_E(t) \exp(-it\varepsilon). \tag{3.9}$$

As is known, only the characteristic function of the form

$$\varphi_X(t) = \exp\left(iat - \frac{\sigma^2 t^2}{2}\right) \tag{3.10}$$

gives rise to the Gaussian distribution function

$$\rho_X(x) = \frac{1}{\sqrt{2\pi\sigma^2}} \exp\left[-\frac{(x-a)^2}{2\sigma^2}\right]. \tag{3.11}$$

Here the parameters a and σ imply, respectively, the mean value and r.m.s. deviation (the variance is σ^2) of random process x. Note that in this case the determinative contribution to the integral (3.9) is given by vicinity of the point $t_0 = -i(x-a)/\sigma^2$ of the width $\Delta t \sim \sqrt{2}/\sigma$.

We substitute the explicit expression (3.1) for $\varepsilon(n)$ into the definition (3.8) for characteristic function $\varphi_E(t)$,

$$\varphi_E(t) \equiv \overline{\exp[it\varepsilon(n)]} = \exp\left\{it\left[\overline{\varepsilon} - \overline{\alpha}\sqrt{\frac{K_\varepsilon(0)}{K_\alpha(0)}} \sum_{n'=-\infty}^{\infty} F(n')\right]\right\}$$

$$\times \overline{\exp\left[it\sqrt{\frac{K_\varepsilon(0)}{K_\alpha(0)}} \sum_{n'=-\infty}^{\infty} F(n')\alpha(n-n')\right]}. \tag{3.12}$$

Then, in Eq. (3.12) we should present the second-line exponential function of the sum of arguments as a product of exponential functions and take into account statistical independence of uncorrelated random variables $\alpha(n)$. This procedure yields

$$\varphi_E(t) = \exp\left\{it\left[\overline{\varepsilon} - \overline{\alpha}\sqrt{\frac{K_\varepsilon(0)}{K_\alpha(0)}} \sum_{n=-\infty}^{\infty} F(n)\right]\right\}$$

$$\times \prod_{n'=-\infty}^{\infty} \overline{\exp\left[it\sqrt{\frac{K_\varepsilon(0)}{K_\alpha(0)}} F(n')\alpha(n-n')\right]}. \tag{3.13}$$

According to definition, the average of the exponential function standing under the sign of the product in Eq. (3.13), reads

$$\overline{\exp\left[it\sqrt{\frac{K_\varepsilon(0)}{K_\alpha(0)}}F(n')\alpha(n-n')\right]} = \int_{-\infty}^{\infty} d\alpha\, \rho_A(\alpha)\exp\left[it\sqrt{\frac{K_\varepsilon(0)}{K_\alpha(0)}}F(n')\alpha\right]$$

$$= 1-\overline{\alpha}\left\{1-\exp\left[it\sqrt{\frac{K_\varepsilon(0)}{K_\alpha(0)}}F(n')\right]\right\} \tag{3.14a}$$

$$\approx \exp\left[it\overline{\alpha}\sqrt{\frac{K_\varepsilon(0)}{K_\alpha(0)}}F(n') - \frac{K_\varepsilon(0)t^2}{2}F^2(n')\right]$$

$$\text{for} \quad \left|t\sqrt{\frac{K_\varepsilon(0)}{K_\alpha(0)}}F(n')\right| \ll 1. \tag{3.14b}$$

The last *asymptotical* expression (3.14b) was obtained from the previous *exact* Eq. (3.14a) with the use of approximate equality

$$1-\overline{\alpha}\,[1-\exp(ix)] \approx \exp\left[i\overline{\alpha}x - \frac{\overline{\alpha}(1-\overline{\alpha})}{2}x^2\right]$$

$$\text{for} \quad |x| \ll 1, \qquad \overline{\alpha}(1-\overline{\alpha}) \lesssim 1; \tag{3.15}$$

and expression for the variance $C_\alpha(0)=\overline{\alpha}(1-\overline{\alpha})$ of the dichotomous white noise $\alpha(n)$. It can be easily ensured that at $|x| \ll 1$ both sides of Eq. (3.15) coincides up to second-order (quadratic) terms in x.

Suppose that the main contribution to the integral of Eq. (3.9) be given by such values of t for which the condition written down in Eq. (3.14b) is fulfilled. In this case we can substitute in Eq. (3.13) the corresponding asymptotic (3.14b). After we take into account the normalization (3.4), the expression for characteristic function $\varphi_E(t)$ gets the desirable form (3.10) with mean value $a=\overline{\varepsilon}$ and variance $\sigma^2 = C_\varepsilon(0)$,

$$\varphi_E(t) = \exp\left(i\overline{\varepsilon}t - \frac{K_\varepsilon(0)t^2}{2}\right). \tag{3.16}$$

As a result, in line with Eqs. (3.10) and (3.11), we arrive at the expected Gaussian probability density $\rho_E(\varepsilon)$ for random sequence $\varepsilon(n)$ with long-range correlations,

$$\rho_E(\varepsilon) = \frac{1}{\sqrt{2\pi K_\varepsilon(0)}}\exp\left[-\frac{(\varepsilon-\overline{\varepsilon})^2}{2K_\varepsilon(0)}\right], \tag{3.17}$$

provided the condition

$$(\varepsilon-\overline{\varepsilon})^2/2K_\varepsilon(0) \ll R_c \tag{3.18}$$

is fulfilled. The deviations from the Gaussian shape may appear only at the far tails, where $(\varepsilon-\overline{\varepsilon})^2/2K_\varepsilon(0) \gg R_c$. Note that for a continuous Gaussian distribution of $\alpha(n)$ Eq.(3.17) is exact for any value of R_c.

From the above consideration it is clear that the correlated sequence $\varepsilon(n)$ may be generated using very different uncorrelated sequences $\alpha(n)$, including the binary white noise. However, if we start with the binary white noise $\alpha(n)$, the generated sequence $\varepsilon(n)$ is obviously a non-binary one since at any site n the value of $\varepsilon(n)$ in Eq. (3.1) results from a linear superposition of binary entries. This means that the direct application of the convolution method does not generate a binary correlated sequence.

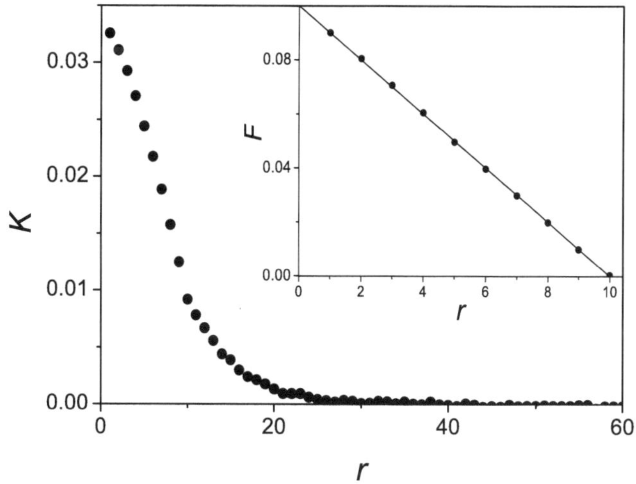

Figure 1 Calculated correlation function $K(r)$ of the Markov chain constructed with the model memory function $F(r)$, Eq. (3.19), shown by the solid line in the inset. Dots in the inset correspond to the memory function reconstructed by solving Eq. (2.109) with correlation function $K(r)$ presented in the main panel.

3.2 MARKOV GENERATION

In Section 2 a general theoretical model of additive Markov chain was considered. In this Section a method of *additive Markov chain* construction is described and examples of numerical construction of such objects is presented.

3.2.1 Numerical reconstruction of the memory function

Let us present the consecutive steps for constructing the random sequence with desired correlation function:

- Assign required average value $\bar{a}$, memory length N and correlation function $K(r)$ for $r = 1, \dots, N$;
- Solving Eq. (2.109), we determine the memory function $F(r)$ and the conditional probability function $P(a_i = 1 \mid T_{N,i})$;
- Construct random correlated sequence a_i with obtained conditional probability function $P(a_i = 1 \mid T_{N,i})$.

Let us demonstrate the robustness of additive Markov chains method by numerical simulations. We consider a model *memory function*,

$$F(r) = 0.1(1 - r/10), \qquad 1 \leqslant r \leqslant 10, \tag{3.19}$$

shown in the inset in Fig. 1 by solid line. Using Eq. (1.2), we construct a random unbiased, $\bar{a} = 1/2$, Markov chain. Then, with the help of the constructed binary sequence of the length 3×10^6, we calculate numerically the correlation function $K(r)$ by solving the set of N linear equations (2.109). The result of these calculations is given in Fig. 1. We can see that the correlation function $K(r)$ mimics roughly the memory function $F(r)$ over the region $1 \leqslant r \leqslant 10$. In the region $r > 10$, the memory function is equal to zero but the correlation function does not vanish (The existence of the "additional tail" in the correlation function is in agreement with Ref. [58] and corresponds to the fact that the correlation length is always larger than the region of memory function action). Then, using the obtained correlation function $K(r)$, we solve numerically Eq. (2.109). The result is shown in the inset in Fig. 1 by dots. We can see an excellent agreement of initial, Eq. (3.19), and reconstructed memory functions $F(r)$.

Thus, relation (2.109) between the correlation and memory functions allows us to construct a binary sequence with a *prescribed correlation function.*

Yet another approach to numerical finding the memory function is an iteration procedure. For its realization let us rewrite Eq. (2.109) in the form,

$$F(r) = \frac{K(r)}{K(0)} - \sum_{r'=1,\ r'\neq r}^{N} \frac{K(r-r')}{K(0)} F(r'). \tag{3.20}$$

Using Eq. (3.20) with starting iteration $F_0(r) = 0$, we obtain the formula,

$$F_{n+1}(r) = \frac{K(r)}{K(0)} - \sum_{r'=1,\ r'\neq r}^{N} \frac{K(r-r')}{K(0)} F_n(r'),\ n \geqslant 0. \tag{3.21}$$

Thus, the memory function can be presented as the series,

$$F(r) = \frac{K(r)}{K(0)} - \sum_{r'\neq r} \frac{K(r-r')K(r')}{K^2(0)} + \sum_{r'\neq r} \sum_{r''\neq r'} \frac{K(r-r')K(r'-r'')K(r'')}{K^3(0)} + \ldots \tag{3.22}$$

Note, that the Markov chain with the definite correlation function $K(r)$ exists if the series (3.22) is convergent and the obtained function implies the probability Eq. (1.2) satisfying the requirement $0 \leqslant P(a_i = 1|T_{N,i}) \leqslant 1$ for arbitrary word $T_{N,i}$. If $\bar{a} = 1/2$ we obtain the restriction, $\sum |F(r)| \leqslant 1$. The sufficient, but not necessary, requirement is $\sum_{r=1}^{N-1} |K(r)| \leqslant 1/12 - |K(N)|/3$.

3.2.2 Numerical reconstruction of the correlation function

Let us demonstrate the ability to construct a binary sequence with an arbitrary *prescribed correlation function* by means of Eq. (2.109). As an example, let us consider the model correlation function,

$$K(r) = 0.1 \frac{\sin(r)}{r}, \tag{3.23}$$

presented by the solid line in Fig. 2. We solve Eq. (2.109) numerically to find the memory function $F(r)$ using this correlation function. The result is presented in inset Fig. 2. Then we construct the binary Markov chain using the obtained memory function $F(r)$. To check up the robustness of the method, we calculate the correlation function $K(r)$ of the constructed chain (the dots in Fig. 2) and compare it with Eq. (3.23). We can see an excellent agreement between the initial and reconstructed correlation functions.

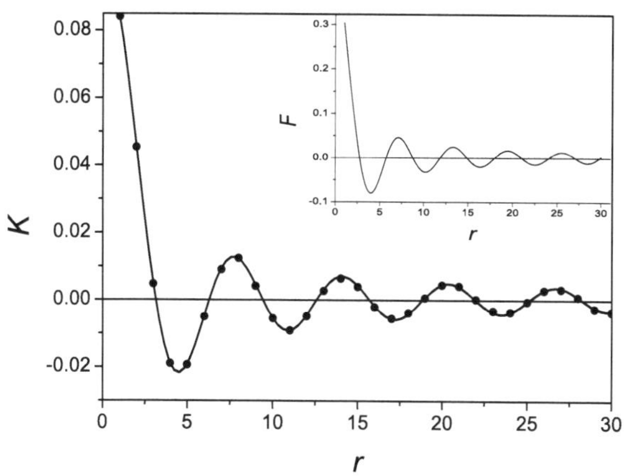

Figure 2 The model correlation function $K(r)$ described by Eq. (3.23) (solid line). The dots correspond to the reconstructed correlation function. In inset, the memory function $F(r)$ obtained by numerical solution of Eq. (2.109) with correlation function Eq. (3.23).

3.2.3 Non-additive generation

As described, the method based on the additive Markov chains appears to be quite suitable for the numerical reconstruction of the sequences with the prescribed correlation function. Here we touch upon the question of the generality of the method.

Consider the 2-step Markov chain of symbols $s_i = \pm 1$ defined by the conditional probability function

$$P(s_i = 1 | s_{i-1}, s_{i-2}) = \frac{1+\varepsilon}{2} + \frac{1-2\varepsilon}{2} s_{i-1} s_{i-2}, \tag{3.24}$$

where positive parameter $\varepsilon < 1/2$.

Now we write the equations for the correlation functions in the way of Subsection 2.3.3 and deduce the recurrence relation for $K(r)$ and its initial conditions:

$$K(r) = \varepsilon(1-2\varepsilon)K(r-2) + (1-2\varepsilon)^2 K(r-3), \quad r = 2, 3, \ldots, \tag{3.25}$$

$$\bar{s} = \frac{1}{3-2\varepsilon}, \quad K(0) = 1 - \bar{s}^2, \quad K(1) = (1-\varepsilon-\bar{s})\bar{s}. \tag{3.26}$$

It can be solved analytically by the characteristic equation,

$$\lambda^3 = \varepsilon(1-2\varepsilon)\lambda + (1-2\varepsilon)^2. \tag{3.27}$$

However, we restrict our study to the approximate value of the correlation function for $\varepsilon \ll 1$ and $r \ll \varepsilon^{-1}$ only,

$$K(r) \simeq \begin{cases} 8/9 - (30r-4)\varepsilon/27, & r = 3k, \\ 2/9 - (6r-7)\varepsilon/27, & r = 3k+1, \\ 2/9 + \varepsilon/27, & r = 3k+2. \end{cases} \tag{3.28}$$

Here $r \geqslant 0$ and $K(-r) = K(r)$.

Let us try and find the additive Markov chain that possesses the same average $\bar{s}$ and pair correlation function (3.28). For this purpose we solve the following system (see Subsection 2.2.1),

$$K(r) = \sum_{r'=1}^{\infty} K(r-r')F(r'), \quad r = 1, 2, \ldots. \tag{3.29}$$

where $F(r)$ is the memory function of the required additive Markov chain described by the conditional probability function,

$$P(s_i = 1|s_{i-1}, s_{i-2}, \ldots) = \frac{1+\bar{s}}{2} + \sum_{r=1}^{\infty} \frac{F(r)}{2}(s_i - \bar{s}) \tag{3.30}$$

We seek for the solution in the main approximation (for $\varepsilon \ll 1$), because this approximation points out the existence of the Markov chain. Namely, if we obtain that

$$Z = \left| \bar{s} \left(1 - \sum_{r=1}^{\infty} F(r) \right) \right| + \sum_{r=1}^{\infty} |F(r)| > 1, \tag{3.31}$$

then it follows that the additive Markov chain cannot possess the presented pair correlation function (3.24) because the required probability function does not belong to interval [0, 1].

When the problem (3.29) being considered in the zero approximation only, we arrive at the following conditions,

$$\sum_{r=1}^{\infty} F_0(3r) = 1, \tag{3.32a}$$

$$\sum_{r=1}^{\infty} F_0(3r-1) = \sum_{r=1}^{\infty} F_0(3r-2) = 0. \tag{3.32b}$$

Here $F_0(r)$ is the main approximation.

Then we proceed to the next approximation and examine the expressions $K(r-3) - 2K(r) + K(r+3)$. For Eq. (3.29) it yields the following recurrence relation

$$F_0(r) + 10F_0(r+1) + F_0(r+2) = 0, \quad r = 3, 4, \ldots. \tag{3.33}$$

Equation (3.33) with equations (3.32) gives us the explicit expression for $F_0(r)$:

$$F_0(1) = 5 - 2\sqrt{6}, \quad F_0(2) = 20\sqrt{6} - 49, \tag{3.34}$$

$$F_0(r) = 18(\sqrt{6}-3)(2\sqrt{6}-5)^r, \quad r = 3, 4, \ldots. \tag{3.35}$$

Thus, we obtain the expression for Z in the main approximation,

$$Z_0 = \frac{3-\sqrt{6}}{\sqrt{6}-2} > 1. \tag{3.36}$$

To find the next approximations of $F(r)$ we should write rather cumbersome equations. The numerical calculations for the next approximations yield

$$Z = \frac{3-\sqrt{6}}{\sqrt{6}-2} + C_1\varepsilon + C_2\varepsilon^2 + \ldots, \tag{3.37}$$

$$C_1 \approx -4.347, \quad C_2 \approx 5.717. \tag{3.38}$$

So, we have shown that there are dichotomous chains, where the average and pair correlation function cannot be expressed in terms of the dichotomous additive Markov chain. Although the additive Markov chains is simpler for both analytical and numerical calculations, we have to establish that the additive method is not general against the correlation function reconstruction.

3.3 FILTERING PROBABILITY FUNCTIONS

3.3.1 Transformation of uncorrelated sequence into correlated one

Let us now consider, Reference [57], the problem of generating a binary sequence $\varepsilon(n)$ with prescribed correlations, assuming that an original sequence under transformation, $\alpha(n)$, is also binary. We suppose that both the sequence $\varepsilon(n)$ and $\alpha(n)$ contain 0's and 1's. Let the nth site for $\varepsilon(n)$ is associated with the number P_n $(0 \leq P_n \leq 1)$, which is the probability of "1" obtaining at this site. In order to calculate the *filtering probabilities* P_n from white noise $\alpha(n)$, in analogy with Eq. (3.1) we propose the linear transformation

$$P_n = \overline{\varepsilon} + \sqrt{\frac{K_\varepsilon(0)}{K_\alpha(0)}} \sum_{n'=-\infty}^{\infty} F(n-n')[\alpha(n') - \overline{\alpha}]. \tag{3.39}$$

Here $K_\varepsilon(0) = \overline{\varepsilon}(1-\overline{\varepsilon})$ and $K_\alpha(0) = \overline{\alpha}(1-\overline{\alpha})$ are the variances of $\varepsilon(n)$ and $\alpha(n)$, respectively, and $F(n)$ is an unknown modulation function to be determined. Having the value of P_n, the nth symbol is generated by drawing randomly a number from the interval [0,1]. If this number is less than P_n, then $\varepsilon(n) = 1$, otherwise, $\varepsilon(n) = 0$. Thus, a binary sequence $\varepsilon(n)$ can be generated, once the set of numbers $\{P_n\}_{n=-\infty}^{\infty}$ is known due to Eq. (3.39). The values of P_n are correlated in the following way,

$$\overline{[P_{n+r} - \overline{\varepsilon}][P_n - \overline{\varepsilon}]} = K_\varepsilon(0) \sum_{n=-\infty}^{\infty} F(n)F(n+r), \qquad \text{for} \quad r \neq 0. \tag{3.40}$$

According to the method of generation of the sequence $\varepsilon(n)$ from the filtering probabilities, the probability of the symbol $\varepsilon(n+r)$ obtaining at the $(n+r)$th site does not depend on the emergence of the symbol $\varepsilon(n)$ at nth site (for $r \neq 0$). Therefore, the product $P_{n+r}P_n$ gives the joint probability of 1 obtaining at the nth and $(n+r)$th sites. If 0 appears at either of these sites, the corresponding pair does not contribute to the product $\overline{\varepsilon(n+r)\varepsilon(n)}$. Hence, the correlation function for sequence $\varepsilon(n)$ can be expressed through the correlation function of the filtering probabilities

$$\overline{[\varepsilon(n+r) - \overline{\varepsilon}][\varepsilon(n) - \overline{\varepsilon}]} = \lim_{N\to\infty} \frac{1}{2N+1} \sum_{n=-N}^{N} P_{n+r}P_n - \overline{\varepsilon}^2 = \overline{[P_{n+r} - \overline{\varepsilon}][P_n - \overline{\varepsilon}]}. \tag{3.41}$$

As we can see, the correlations in the sequence $\varepsilon(n)$ occur because of the correlations between the filtering probabilities. The latter are enforced by the modulation function, see Eq. (3.39). Thus, using Eqs. (3.40) and (3.41), the relation between the correlator of the binary sequence and the modulation function can be written as

$$\mathcal{K}_\varepsilon(r) = \sum_{n=-\infty}^{\infty} F(n)F(n+r) \qquad \text{for} \quad r \neq 0\,; \tag{3.42}$$

$$\mathcal{K}_\varepsilon(0) = 1. \tag{3.43}$$

The normalization condition (3.43) is the property of the correlator $\mathcal{K}_\varepsilon(r)$. It should be stressed that unlike Eq. (3.3) that is valid for all values of r including $r = 0$, in the case of binary sequences the derived Eq. (3.42) is not valid for $r = 0$. Therefore, the sum $\sum_n F^2(n)$ remains undefined and has to be considered as a free constant,

$$A = \sum_{n=-\infty}^{\infty} F^2(n) = \frac{1}{2\pi}\int_{-\pi}^{\pi} \hat{\mathcal{F}}^2(k)dk. \tag{3.44}$$

This constant appears now in the Fourier transform of Eq. (3.42) as follows,

$$\hat{\mathcal{K}}_\varepsilon(k) = 1 - A + \mathcal{F}^2(k) \tag{3.45}$$

Using Eq. (3.45) the relation between the modulation function and the pair correlator of the binary sequence can be written in the form,

$$F(n) = \frac{1}{\pi}\int_0^{\pi} \left[\hat{\mathcal{K}}_\varepsilon(k) - 1 + A\right]^{1/2} \cos(kn)dk. \tag{3.46}$$

Thus, Eqs. (3.39) and (3.46) define the algorithm of the generation of a binary sequence with the prescribed pair correlator.

The values of P_n in Eq. (3.39) can be associated with probability if $0 \le P_n \le 1$. This condition leads to the following constraints for the unknown constant A,

$$\sum_{n=-\infty}^{\infty} |F(n)| \le \frac{\mathtt{min}(\overline{\varepsilon}, 1-\overline{\varepsilon})}{\sqrt{\overline{\varepsilon}(1-\overline{\varepsilon})}} \, \frac{\sqrt{\overline{\alpha}(1-\overline{\alpha})}}{\mathtt{max}(\overline{\alpha}, 1-\overline{\alpha})} \le 1. \tag{3.47}$$

Taking into account that the argument of the square-root in Eq. (3.46) must be positive, this inequality can be rewritten (for $\bar{\alpha} = \bar{\varepsilon} = 1/2$) in terms of the constant A,

$$1 - \hat{\mathcal{K}}_\varepsilon(k) \le A = \sum_{n=-\infty}^{\infty} F^2(n) \le \sum_{n=-\infty}^{\infty} |F(n)| \le 1\,. \tag{3.48}$$

The constraints imposed by Eqs. (3.47) and (3.48) limit the class of generated binary sequences that can have a given pair correlator. In particular, a binary sequence with the slow-decaying correlator $\mathcal{K}_\varepsilon(r) = \sin(ar)/ar$ cannot be generated with the use of the proposed algorithm since the sum $\sum_n |F(n)|$ diverges in this case. As is known, the power-decaying correlators provide an emergence of a kind of mobility edges in systems with random one-dimensional potentials [25, 72]. In this case, a sharp transition from localized to delocalized eigenstates occurs when crossing some value in the energy spectrum, specified by the mobility edge. Although such mobility edges were experimentally observed for a continuous distribution of fluctuations in artificially fabricated site-potentials with long-range correlations [70], it is not clear yet, if they do exist for a correlated binary sequence. That is why the existence of a mobility edge in a sequence of nucleotides in a DNA molecule is still questionable [73, 74].

Note that the sequences with monotonously power-law decaying correlations are used for modelling of superdiffusion [22], and anticorrelations (i.e. a long negative tail for $\mathcal{K}_\beta(r)$) are necessary in the fluctuations of the force, driving a subdiffusion process, Ref. [26].

Thus, the proposed method can be used for generation of binary sequences with long-range correlations. To the best of our knowledge, it is the first successful attempt to generate a random binary sequence with power-decaying correlations from white noise. It is worth mentioning that a *renewal* binary sequence with prescribed correlations can be generated using the relation $d^2\mathcal{K}_\varepsilon(n)/dn^2 = \psi(n)$ between the correlation function $\mathcal{K}_\varepsilon(n)$ and the probability density $\psi(n)$ of occurrence of a laminar layer of length n, Ref. [76]. The binary sequences, which can be designed from this relation are those with $\mathcal{K}''_\varepsilon(n) > 0$. This constraint does not allow generation of the sequences with alternating-sign correlations. Numerous applications of the aforementioned relation for stochastic dynamical processes with intermittency can be found in Reference [77].

In comparison to a sequence of continuously distributed random elements, a binary sequence leaves much less freedom to implant correlations. Therefore the proposed modification of the convolution method allows the generation of random binary sequences with a much narrower class of correlation functions than the traditional convolution method. In the latter case practically any function, decaying at $r \to \infty$ and possessing nonnegative Fourier transformation (see Eq. (3.5)), can serve as $\mathcal{K}_\beta(r)$ - a correlator of the random sequence (3.1). More examples of sequences with short- and long-range correlations generated by the convolution method can be found in References [22, 25, 26, 70, 72].

3.3.2 Infinity-step filtering

It is difficult to hope that by one step of generation starting from *uncorrelated* sequence we can construct *strongly correlated* one. For this reason we consider *a chain of transformations* sequences, [78]. For this we have to construct the transformation of a *correlated* sequence into another correlated sequence. After introducing a new notation $\mathcal{K}^{(m)}(r)$ for the normalized correlation function $\mathcal{K}_\varepsilon(r)$, constructed after m-step of filtering and $\mathcal{K}^{(m-1)}(r)$ instead of $\mathcal{K}_a(r)$, we get,

$$\mathcal{K}^{(m)}(r) = \sum_{n,n'=-\infty}^{\infty} F(n)F(n')\mathcal{K}^{(m-1)}(r+n-n') + \left[1 - \sum_{n,n'=-\infty}^{\infty} F(n)F(n')\mathcal{K}^{(m-1)}(n-n')\right]\delta_{r,0}\,. \tag{3.49}$$

Fourier transformation of Eq. (3.49) reads

$$\hat{\mathcal{K}}^{(m)}(k) = \mathcal{F}^2(k)\hat{\mathcal{K}}^{(m-1)}(k) + B^{(m-1)}. \tag{3.50}$$

Supposing existence of "fixed points" $\mathcal{K}(r) = \lim_{m\to\infty} \mathcal{K}^{(m)}(r)$ of transformations (3.49) and $\hat{\mathcal{K}}(k) = \lim_{m\to\infty} \hat{\mathcal{K}}^{(m)}(k)$ of Eq. (3.50) we have

$$\hat{\mathcal{K}}(k) = \mathcal{F}^2(k)\hat{\mathcal{K}}(k) + B. \tag{3.51}$$

The condition of $\hat{\mathcal{K}}^{(m)}(k)$ convergence to the solution of Eq. (3.51) following from Eq. (3.50), is

$$|\mathcal{F}(k)| < 1. \tag{3.52}$$

It follows from Eq. (3.48) and the chain of inequalities,

$$|\mathcal{F}(k)| = |\sum_{r=-\infty}^{\infty} F(r)\exp(ikr)| \leq \sum_{n=-\infty}^{\infty} |F(n)| < 1, \tag{3.53}$$

that the convergency condition Eq. (3.52) is always fulfilled.

More convenient but equivalent to Eq. (3.51) definition for B expressed in the limit $m \to \infty$ by means of Fourier transform is,

$$B = \frac{1}{2\pi}\int_{-\pi}^{\pi} dk\hat{\mathcal{K}}(k)[1 - \mathcal{F}^2(k)]. \tag{3.54}$$

The method of filtering probability, suggested here, makes it possible to solve two different kinds of problems. The first one, the direct problem, is to construct a real sequence and find a correlation function corresponding to the given modulation function $F(n)$. The second one, the inverse problem, supposes a reconstruction of $F(n)$ via prescribed $\mathcal{K}(r)$.

The solution of Eq. (3.51) considered as the direct problem of finding the sequence possessing given $F(r)$ is

$$\mathcal{K}(r) = I(r)/I(0), \tag{3.55}$$

$$I(r) = \frac{1}{2\pi}\int_{-\pi}^{\pi} dk \frac{\cos(kr)}{1 - \mathcal{F}^2(k)}. \tag{3.56}$$

How to solve the inverse problem? Although the solution of the direct problem does not depend explicitly on the constant B, we cannot avoid it solving the inverse problem. From Eq. (3.51) we obtain

$$F(r) = \frac{1}{2\pi}\int_{-\pi}^{\pi} dk \cos(kr)\sqrt{1 - \frac{B}{\hat{\mathcal{K}}(k)}}. \tag{3.57}$$

In reality the value of B determines the rate of convergence but not the solution itself. The B has to be chosen to satisfy the inequality Eq. (3.53).

In both cases of the direct and inverse problems when we construct a random sequence, it is natural to choose as the starting point an uncorrelated homogeneous stationary random sequence with $\mathcal{K}(r) = \delta_{r,0}$. Taking into account n independence of $F(n')$ in Eq. (3.39) we obtain at the end after the chain of transformations the strongly stationary binary chain.

3.3.3 Examples

As an example, of short-range correlations let us consider the exponential binary correlator with the corresponding Fourier representation,

$$\mathcal{K}_\varepsilon(r) = \exp(-\gamma|r|), \quad \hat{\mathcal{K}}_\varepsilon(k) = \frac{\sinh\gamma}{\cosh\gamma - \cos k}. \tag{3.58}$$

Since $\hat{\mathcal{K}}_\varepsilon(k)$ reaches its minimum at $k = \pi$, the condition $1 - \hat{\mathcal{K}}_\varepsilon(k) \le A$ is satisfied for $A = 1 - \hat{\mathcal{K}}_\varepsilon(\pi) = 1 - \tanh(\gamma/2)$. Therefore, we have,

$$\mathcal{F}^2(k) = \frac{1 + \cos k}{\cosh\gamma - \cos k} \tanh\frac{\gamma}{2}, \tag{3.59}$$

and the function $F(n)$ reads

$$F(n) = \frac{1}{\pi}\sqrt{\tanh\frac{\gamma}{2}} \int_0^\pi \cos(kn)\sqrt{\frac{1+\cos k}{\cosh\gamma - \cos k}}\, dk\,. \tag{3.60}$$

For $n \gg 1$ the modulation function decays as follows,

$$F(n) \approx \frac{(-1)^{n+1}}{2n^2\cosh(\gamma/2)}\sqrt{\tanh\frac{\gamma}{2}} \propto 1/n^2\,, \tag{3.61}$$

therefore, the sum $\sum_n |F(n)|$ converges. It, however, exceeds 1 for $\gamma < \gamma_{cr} \approx 1.60$, thus violating the last inequality in Eq. (3.48). The numerical simulation shows that for $\gamma > \gamma_{cr}$ the method works quite well, giving a possibility to design binary chains with exponentially decaying correlations.

Since a method of generation of a binary sequence with exponentially correlations has already been reported [75], it is important to demonstrate the validity of the proposed algorithm for generation of binary sequences with long-range correlations. Let us consider power-lay decaying correlation function,

$$\mathcal{K}_\varepsilon(n) = \frac{(\pm 1)^n}{n^2} a, \quad n \neq 0. \tag{3.62}$$

For these correlators the modulation function and its Fourier transform can be explicitly calculated

$$F(n) = \pm\sqrt{\frac{a}{2}}\frac{1}{\pi n^2}\left[1 - (-1)^n\right], \quad F(0) = \frac{\pi}{2}\sqrt{\frac{a}{2}}, \tag{3.63}$$

$$\hat{F}(k) = \sqrt{\frac{a}{8}}\left(\pi \mp (2|k| - \pi)\right), \quad -\pi \leqslant k \leqslant \pi. \tag{3.64}$$

The constant a is obtained from Eq. (3.48), which leads to the following inequality,

$$\sum_{n=-\infty}^{\infty} |\, F(n) \,| = \pi\sqrt{\frac{a}{2}} \le 1. \tag{3.65}$$

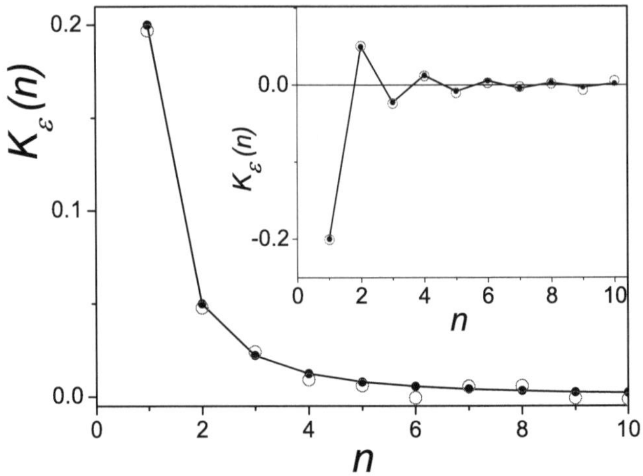

Figure 3 Correlation function of the binary sequence with power-law decaying correlations. The inset shows the correlation function with alternating-sign. Solid circles are for the theoretical formulas Eq. (3.62) and open circles are for the numerical data for generated sequences of the length 10^5. The value $\mathcal{K}(0) = 1$ is not shown.

It is satisfied for $0 < a \leq 2/\pi^2 \approx 0.203$. For numerical calculations we chose $a = 0.2$. Two filtering functions Eq. (3.63) were used to generate two binary sequences of the length 10^5 each. The correlators of these sequences are plotted in Fig. 3. A good agreement between theoretical Eq. (3.62) and numerical results is obtained for the sequence with monotonically decaying correlations and for the sequence with alternating-sign correlations.

Let us calculate the correlation function of the sequence obtained after the infinity step filtering with the modulation function define by Eq. (3.63). Using Eqs. (3.55) and (3.56), and integrating by parts we arrive at the following asymptotic expression for the correlation function,

$$K^{(\infty)}(n) \approx \frac{(\pm 1)^n}{n^2} \pi a \sqrt{2a} \left(1 - \pi^2 a/2\right)^{-2} \ln^{-1} \left[\frac{1 + \pi\sqrt{a/2}}{1 - \pi\sqrt{a/2}} \right], \quad n \gg 1. \tag{3.66}$$

Comparing the last equation with Eq. (3.62) we can see the increased correlation "strength":

$$\frac{K^{(\infty)}(n)}{K^{(1)}(n)} \approx \pi\sqrt{2a}(1 - \pi^2 a/2)^{-2} \ln^{-1} \left[\frac{1 + \pi\sqrt{a/2}}{1 - \pi\sqrt{a/2}} \right] > 1. \tag{3.67}$$

So, we can presume that the infinity step filtering produced the stronger correlations than one step procedure.

3.4 SIGNUM-GENERATION

In this Subsection we analyze the signum-generation method for creating random dichotomic sequences with prescribed correlation properties. The method is based on a binary mapping of the convolution of continuous random numbers with some function originated from the Fourier transform of a binary correlator. The goal of our study is to reveal conditions under which we can construct binary sequences with a given pair correlator.

A direct way to create dichotomic correlated sequences is to apply the signum function to the sequence of continuous values, obtained with the convolution method. This method is based on the convolution of a white-noise sequence with some function that is determined by the desired pair correlator. However, as was numerically found in [73], the created binary sequence turns out to have the pair correlator different from the expected one. To date, it remains unclear how to construct binary sequences having the same pair correlators as in the sequences with continuous values of their elements. In spite of its quite simple form, the signum function method is not rigorously analyzed in the literature.

3.4.1 Binary versus Gaussian

Now we analyze the construction of a dichotomic sequence γ_n by means of the *signum function generation* (SFG) method. It uses an intermediate correlated disorder β_n obtained as a convolution of the uncorrelated Gaussian noise α_n and modulation function $G(n)$. Specifically, the γ-sequence is defined by

$$\gamma_n = \text{sign}(\beta_n), \tag{3.68a}$$

$$\beta_n = \overline{\beta} + \sum_{n'=-\infty}^{\infty} G(n-n')\alpha_{n'}. \tag{3.68b}$$

The initial Gaussian white-noise chain consists of stochastic variables α_n with the zero mean, unit variance and corresponding probability density. Respectively,

$$\overline{\alpha} = 0, \qquad \overline{\alpha_n \alpha_{n'}} = \delta_{n,n'}, \tag{3.69a}$$

$$\rho_A(\alpha_n = \alpha) = \frac{1}{\sqrt{2\pi}} \exp(-\alpha^2/2). \tag{3.69b}$$

The bar over a random symbol or function implies the stochastic average.

Evidently, the constructed dichotomic sequence γ_n does not change if we normalize the intermediate sequence β_n by an arbitrary factor. Hence, without any loss of generality, we can admit the variance of β_n be equal to unity. Then, $\overline{\beta}$ is the mean value of β_n and its correlation function $K_\beta(r)$,

$$K_\beta(r) = \overline{(\beta_n - \overline{\beta})(\beta_{n+r} - \overline{\beta})}, \tag{3.70}$$

is normalized to unity, $K_\beta(0) = 1$. By direct substitution of Eq. (3.68b) into definition (3.70), the correlator $K_\beta(r)$ is readily associated with the modulation function $G(n)$,

$$K_\beta(r) = \sum_{n=-\infty}^{\infty} G(r-n)G(n). \tag{3.71}$$

Owing to the evenness of $K_\beta(r) = K_\beta(-r)$ and in accordance with Eq. (3.71), we can also restrict the function $G(n)$ to the class of even functions, $G(-n) = G(n)$. Note that the condition $K_\beta(0) = 1$ gives rise to the following normalization for $G(n)$,

$$K_\beta(0)=\sum_{n=-\infty}^{\infty} G^2(n)=1. \tag{3.72}$$

It is convenient to pass to the Fourier transform of Eq. (3.71) with the use of the standard expressions (see Eqs. (2.122)–(2.124)).

The Fourier transform $\hat{K}_\beta(k)$ of the pair correlator $K_\beta(r)$ is known as the *power spectrum* of random β-chain. The correlator $K_\beta(r)$ is non-negative function of the wave number k. Analogously to Eq. (2.122), for the modulation function $G(n)$ we can define its Fourier transform $\hat{G}(k)$, which is also real and even function, $\hat{G}(-k)=\hat{G}(k)$.

The Fourier representation of Eq. (3.71) reads

$$\hat{K}_\beta(k)=\hat{G}^2(k). \tag{3.73}$$

Thus, we arrive at the following expression for the modulation function $G(n)$,

$$G(n)=\frac{1}{\pi}\int_0^\pi dk\, \hat{K}_\beta^{1/2}(k)\cos(kn). \tag{3.74}$$

Evidently, the solution (3.74) automatically satisfies the normalization condition (3.72).

Since the initial chain α_n is a delta-correlated Gaussian noise, the intermediate variables β_n also constitute a Gaussian random sequence with single-point distribution function $\rho_B(\beta)$ (see Appendix C),

$$\rho_B(\beta_n=\beta)=\frac{1}{\sqrt{2\pi}}\exp[-(\beta-\overline{\beta})^2/2]. \tag{3.75}$$

In order to reveal statistical properties of the signum-generated dichotomic sequence γ_n, we should associate its mean value $\overline{\gamma}$, variance $C_\gamma(0)$ and pair correlator $C_\gamma(r)$ with the corresponding independent characteristics, namely, the mean value $\overline{\beta}$ and the modulation function $G(n)$ [or, the same, with the intermediate correlator $K_\beta(r)$]. According to the definition of average, we can express the mean value $\overline{\gamma}$ in terms of $\overline{\beta}$ via the error function [79],

$$\overline{\gamma}\equiv\overline{\gamma_n}=\int_{-\infty}^{\infty} d\beta\,\rho_B(\beta)\,\mathrm{sign}(\beta)=\sqrt{\frac{2}{\pi}}\int_0^{\overline{\beta}} dx\exp(-x^2/2)\equiv\mathrm{erf}(\overline{\beta}/\sqrt{2}). \tag{3.76}$$

Similarly to Eq. (2.172), the variance $C_\gamma(0)$ is written in the form

$$C_\gamma(0)\equiv\overline{\gamma_n^2}-\overline{\gamma}^2=1-\overline{\gamma}^2. \tag{3.77}$$

An important characteristic of the stochastic sequence γ_n is the correlation function $C_\gamma(r)$,

$$C_\gamma(r)\equiv\overline{\gamma_n\gamma_{n+r}}-\overline{\gamma}^2=C_\gamma(0)K_\gamma(r). \tag{3.78}$$

Its calculation is performed in Appendix D. Here we write down only the final equation that relates the correlator $K_\gamma(r)$ to $K_\beta(r)$,

$$(1-\overline{\gamma}^2)K_\gamma(r)=\frac{2}{\pi}\int_0^{K_\beta(r)}\frac{dx}{\sqrt{1-x^2}}\exp\left(-\frac{\overline{\beta}^2}{1+x}\right). \tag{3.79}$$

Note that r.h.s. of the latter equation is not elementary function, therefore its analytical study is not simple. However, the case $\overline{\beta}=0$ allows us to perform complete analytical analysis.

3.4.2 Unbiased sequence

If the mean value of intermediate chain β_n vanishes, $\overline{\beta}=0$, then due to Eq. (3.76), the mean value of generated sequence γ_n vanishes also, $\overline{\gamma}=0$. In this case the relation (3.79) turns out to be remarkably simplified,

$$K_\gamma(r) = \frac{2}{\pi}\int_0^{K_\beta(r)} \frac{dx}{\sqrt{1-x^2}} = \frac{2}{\pi}\arcsin[K_\beta(r)]. \tag{3.80}$$

Another equivalent form is

$$K_\beta(r) = \sin\left[\frac{\pi}{2}K_\gamma(r)\right]. \tag{3.81}$$

From the above relations we can conclude that the γ-sequence generated with the discussed signum function method, is random. Indeed, the decay of correlations with an increase of $|r|$ in the intermediate β-chain, also results in the decay of correlations in the generated dichotomic γ-sequence.

Now it is suitable to rewrite Eq. (3.81) in the Fourier representation,

$$\hat{K}_\beta(k) = \mathcal{S}\{K_\gamma\}(k). \tag{3.82}$$

Here the symbol $\mathcal{S}\{\cdot\}(k)$ stands for the operator that transforms the function $K(r)$ by the following rule,

$$\mathcal{S}\{K\}(k) \equiv \sum_{r=-\infty}^{\infty} \sin\left[\frac{\pi}{2}K(r)\right]\exp(-ikr) \tag{3.83a}$$

$$= \left(1-\frac{\pi}{2}\right) + \frac{\pi}{2}\hat{K}(k) + 2\sum_{r=1}^{\infty}\left\{\sin\left[\frac{\pi}{2}K(r)\right] - \frac{\pi}{2}K(r)\right\}\cos(kr). \tag{3.83b}$$

It is important to note that the series (3.83a) can converge very slowly. Therefore, in the analytical and numerical analysis we have to take into account a lot of terms in the sum in order to obtain correct result. To avoid this problem, we have used the following trick that is based on the second equality (3.83b). Namely, since $K(r)\to 0$ when $|r|\to\infty$, the latter sum converges quite rapidly according to the asymptotic relation

$$\sin\left[\frac{\pi}{2}K(r)\right] - \frac{\pi}{2}K(r) \to \frac{\pi^3}{48}K^3(r), \qquad |r|\to\infty. \tag{3.84}$$

The substitution of Eq. (3.82) into Eq. (3.74) yields the following final relation between the modulation function $G(n)$ and the correlator $K_\gamma(r)$ of the generating dichotomic noise γ_n,

$$G(n) = \frac{1}{\pi}\int_0^{\pi} dk\,\sqrt{\mathcal{S}\{K_\gamma\}(k)}\,\cos(kn). \tag{3.85}$$

Since the correlator $K_\gamma(r)$ is supposed to be known, the relation (3.85) should be regarded as the expression determining the modulation function $G(n)$.

As we can see, the SFG method for constructing the correlated dichotomic sequence γ_n with the zero mean, unit variance and prescribed two-point correlator $K_\gamma(r)$ reduces to the following steps. First, starting from a desirable profile of $K_\gamma(r)$ and employing Eqs. (3.83) and (3.85), we have to obtain the modulation function $G(n)$. After, the correlated sequence γ_n can be generated in accordance with Eq. (3.68). However, it is important to take into account the restriction that directly follows from Eq. (3.82). Specifically, since the power spectrum of any random process, in particular $\hat{K}_\beta(k)$, is a non-negative function of the wave number k, the function $\mathcal{S}\{K_\gamma\}(k)$ also has to be non-negative,

$$\mathcal{S}\{K_\gamma\}(k) \geq 0 \qquad \text{for} \qquad |k| \leqslant \pi. \tag{3.86}$$

This condition becomes apparent from Eq. (3.85), in which the function $\mathcal{S}\{K_\gamma\}(k)$ enters as a radicand. In other words, with this method the function $K_\gamma(r)$ can be considered as a correlator of a dichotomic random sequence γ_n, if and only if $\mathcal{S}\{K_\gamma\}(k)$ is a non-negative function of k.

In view of the revealed restriction (3.86), let us make a qualitative analysis of Eq. (3.83b). Note that the first summand $(1-\pi/2)$ is negative. Therefore, if the spectrum $\hat{K}(k)$ vanishes within some interval of k and the whole function $\mathcal{S}\{K\}(k)$ is non-negative there, consequently, the third summand is positive, exceeding the value $(\pi/2-1)$. However, in the third summand only first few terms determine its sign and give correct estimate of the magnitude. Thus, we may expect that the correlation function of a dichotomic chain, which spectrum is zero in some interval, cannot be generated by the SFG method.

It should be stressed that the condition (3.86) is a quite serious restriction to the type of correlators allowed for dichotomic sequences. As a demonstration, let us consider simple example of the one-step additive Markov chain of variables $\varepsilon_n = \{0, 1\}$ obtained according to the conditional probability

$$P(\varepsilon_n = 1|\varepsilon_{n-1}) = \bar{\varepsilon} + (\varepsilon_{n-1} - \bar{\varepsilon}) \exp(-k_c), \tag{3.87}$$

where the parameter k_c is the inverse correlation length. As is known [80], this chain has an exponential pair correlator, $K_{\exp}(r) = \exp\big(-k_c|r|\big)$. It is instructive that the function $\mathcal{S}\{K_{\exp}\}(k)$ takes negative values for small k_c below $k_c^* \approx 1.099\ldots$, see the data in Fig. 4. Thus, the SFG method can reproduce exponential correlator only for $k_c \geqslant k_c^*$, however, not for small k_c in the most interesting region of long-range correlations.

3.4.3 Long-range correlators with step-wise spectrum

Maximal jump. Here we demonstrate that the discussed method cannot be applied for a construction of dichotomic sequences with long-range correlators resulting in the step-wise power spectrum

$$K_\gamma(r) = \frac{\sin(k_c r)}{k_c r}, \tag{3.88a}$$

$$\hat{K}_\gamma(k) = \frac{\pi}{k_c}\Theta(k_c - |k|), \quad 0 < k_c \leqslant \pi, \quad |k| \leqslant \pi. \tag{3.88b}$$

This kind of correlations is of specific interest in view of applications to 1D disordered superlattices with a selective transport, see, e.g. [25, 23]. Here k_c is the correlation parameter (inverse correlation length) to be specified, and $\Theta(x)$ implies the Heaviside unit-step function, $\Theta(x<0)=0$ and $\Theta(x>0)=1$. First, we analyze the values of k_c within the interval

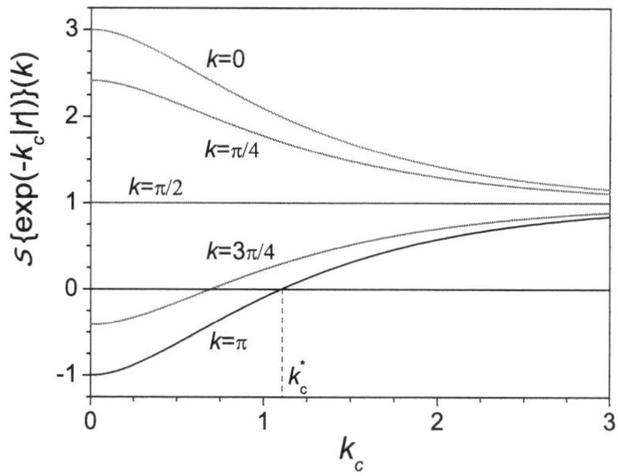

Figure 4 Dependence $\mathcal{S}\{\exp(-k_c|r|)\}(k)$ on the correlation parameter k_c for several values of wave number k.

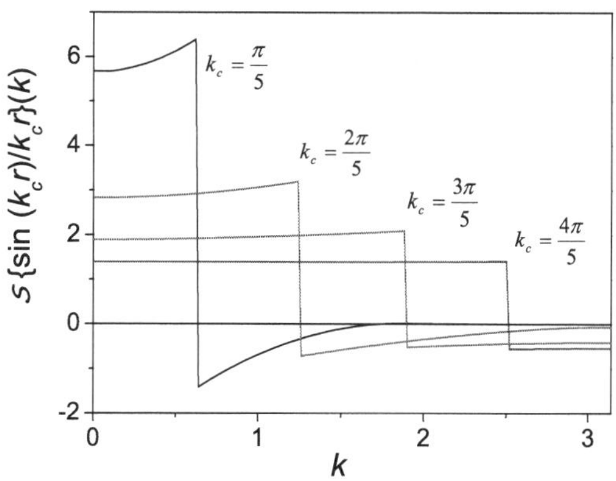

Figure 5 Dependence $\mathcal{S}\{\sin(k_c r)/k_c r\}(k)$ on k for different values of k_c.

$0 < k_c < \pi$. Two specific cases of $k_c = 0$ and $k_c = \pi$ will be considered afterwards.

In the analysis of the SFG method the crucial characteristic is the function $\mathcal{S}\{K_\gamma\}(k)$ defined by Eq. (3.83). It has to be non-negative for any value of the argument k within the interval $|k| \leqslant \pi$, see Eq. (3.86). In the case of the long-range correlator (3.88a) it is suitable to use the following explicit expression for $\mathcal{S}\{K_\gamma\}(k)$,

$$\mathcal{S}\left\{\frac{\sin(k_c r)}{k_c r}\right\}(k) = \left(1 - \frac{\pi}{2}\right) + \frac{\pi^2}{2k_c}\Theta(k_c - |k|) + 2\sum_{r=1}^{\infty}\left\{\sin\left[\frac{\pi}{2}\frac{\sin(k_c r)}{k_c r}\right] - \frac{\pi}{2}\frac{\sin(k_c r)}{k_c r}\right\}\cos(kr). \tag{3.89}$$

It is remarkable that the summand in the last term behaves as $\pi^3/24k_c^3 r^3$ when $r \to \infty$. Hence, at finite k_c the sum converges quite rapidly and *uniformly*. Therefore, the sum is a *continuous* function of k, in particular, at $k = k_c$, and can be easily calculated numerically.

In Fig. 5 the behavior of the radicand (3.89) in Eq. (3.85) is shown for several values of the inverse correlation length k_c. Since $\mathcal{S}\{\cdot\}(k)$ is an even function of the wave number

k, the discussion can be restricted by the interval $0 \leqslant k \leqslant \pi$. From Eq. (3.89) and Fig. 5 we can draw the following conclusions.

1. Due to the last term in expression (3.89) the function $\mathcal{S}\big\{\sin(k_c r)/k_c r\big\}(k)$ increases with an increase of k for all k_c within both intervals $(0, k_c)$ and (k_c, π).
2. The negative jump of $\mathcal{S}\{\sin(k_c r)/k_c r\}(k)$ occurs at $k = k_c$, at the same point where the power spectrum (3.88b) has a jump. The maximal and minimal values of the function are achieved at $k = k_c - 0$ and $k = k_c + 0$, respectively. The jump is exclusively formed by the second term in Eq. (3.89). Therefore, its value reads

$$\mathcal{S}\{\sin(k_c r)/k_c r\}(k = k_c - 0) - \mathcal{S}\{\sin(k_c r)/k_c r\}(k = k_c + 0) = \pi^2/2k_c. \quad (3.90)$$

3. The minimal value $\mathcal{S}\big\{\sin(k_c r)/k_c r\big\}(k = k_c + 0)$ is negative for all finite values of k_c within the interval $0 < k_c < \pi$. The complete dependence of positive function $-\mathcal{S}\big\{\sin(k_c r)/k_c r\big\}(k_c + 0)$ on k_c is depicted in Fig. 6.

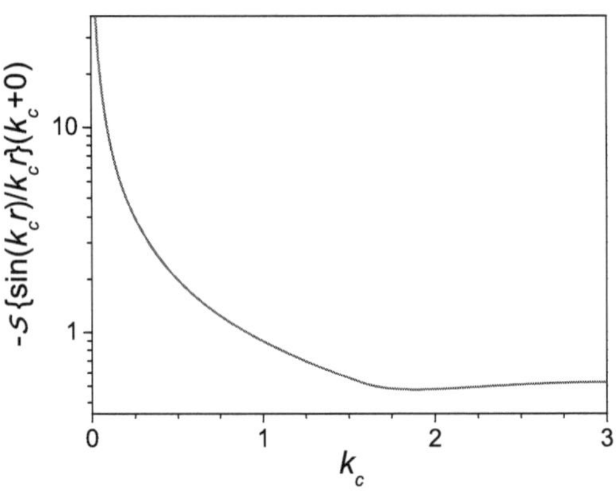

Figure 6 Dependence $-\mathcal{S}\big\{\sin(k_c r)/k_c r\big\}(k_c + 0)$ vs k_c in log-scale.

The third conclusion (above) displays clearly that for all finite values $k_c < \pi$ the correlator and corresponding power spectrum (3.88) cannot be created by making use of the discussed method. Indeed, the modulation function $G(n)$ turns out to be of complex value, see Eq. (3.85).

Now we determine the values of k_c for which the function (3.88a) can not be the pair correlator of a dichotomic sequence regardless of the method of generation. This can be done with the use of the necessary conditions of Subsection 2.3.1.

First, we demonstrate analytically that this function cannot be the correlation function of a dichotomic sequence s_n with arbitrary mean value $\bar{s}$ for small but finite values of k_c, namely, for $0 < k_c \ll 1$. To this end, we take Eq. (2.184) at $r' = 1$ and $r = r_a = [a/k_c]$, where $[x]$ is the integer part of x and a is a constant,

$$\left| \frac{\sin k_c}{k_c} - \frac{\sin k_c(r_a - 1)}{k_c(r_a - 1)} \right| + \frac{\sin k_c r_a}{k_c r_a} \leqslant 1. \quad (3.91)$$

Being expanded in small parameter k_c, this condition reads

$$\left(\frac{\cos a}{a} - \frac{\sin a}{a^2}\right) k_c + O(k_c^2) \leqslant 0. \tag{3.92}$$

It is evident that the l.h.s. of Eq. (3.92) can be positive at some values of a (e.g., for $a = 2\pi + \pi/4$). Thus, the requirement (2.184) is violated.

The inequality (2.186) can also result in new restrictions for allowed values of k_c for the case $\overline{s} = 0$. Rewriting it at $r = 2$ and $r' = 1$, we get

$$4\sin k_c - \sin 2k_c \leqslant 2k_c. \tag{3.93}$$

Numerical analysis shows that this condition does not hold true for all finite k_c from the interval $0 < k_c < k^*$, where $k^* \approx 2.139\ldots$. It is met only at $k_c > k^*$. However, since Eq. (2.186) is just necessary condition, one cannot guarantee an existence of the correlator (3.88a) even at $k_c > k^*$.

At the critical point $k_c = \pi$ the correlator and spectrum (3.88) reduce to $K_\gamma(r) = \delta_{r,0}$ and $\hat{K}_\gamma(k) = 1$. This gives rise to the relations, $\mathcal{S}\{\delta_{r,0}\}(k) = 1$ and $G(n) = \delta_{n,0}$, hence, $\beta_n = \alpha_n$. Consequently, we can apply the SFG. However, this specific case of $k_c = \pi$ is not interesting because from the Gaussian white-noise α_n-sequence the method fabricates the dichotomic chain γ_n again of white-noise type.

As to the singular point $k_c = 0$, here we have $K_\gamma(r) = 1$ for the correlator, and $\hat{K}_\gamma(k) = 2\pi\delta(k)$ for the power spectrum, therefore, the radicand is $\mathcal{S}\{1\}(k) = 2\pi\delta(k)$. We can see that the SFG is formally applicable. Besides, the necessary conditions (3.92) and (3.93) are satisfied automatically. However, this case is a singular one since for any arbitrarily small but finite values of k_c it is not possible to create a dichotomic sequence with the correlation properties (3.88) neither by the SFG or by any other method.

Partial jump. Now we extend our analysis to a more general correlation function that may have various applications. This function also results in a step-wise power spectrum, however, with an additional parameter h that determines the height of step,

$$K_{\gamma,h}(r) = h\,\delta_{r,0} + (1-h)\frac{\sin(k_c r)}{k_c r}\,, \tag{3.94a}$$

$$\hat{K}_{\gamma,h}(k) = h + (1-h)\frac{\pi}{k_c}\Theta(k_c - |k|) > 0\,, \tag{3.94b}$$

$$0 \leqslant h \leqslant 1, \qquad 0 < k_c \leqslant \pi, \quad |k| \leqslant \pi.$$

Eq. (3.94) coincides with Eq. (3.88) if the step-parameter $h = 0$. Otherwise, when $h = 1$ the generated γ-sequence turns into a dichotomic white noise independently of k_c. Also, γ_n becomes delta-correlated at $k_c = \pi$ for arbitrary h. Therefore, at $k_c = \pi$ the conclusions of the previous paragraph are also valid.

In what follows, it is convenient to analyze finite values of $k_c < \pi$. The power spectrum (3.94b) is an even function of the wave number k and has two symmetric jumps at the points $k = \pm k_c$. For positive $0 < k \leqslant \pi$ the spectrum abruptly falls down at $k = k_c$ from the maximal value $\hat{K}_{\gamma,h}(k < k_c) = h + (1-h)\pi/k_c$ to the minimal one, $\hat{K}_{\gamma,h}(k > k_c) = h$. Evidently, this jump can be regarded as a mobility edge of disordered 1D conductors, if the maximal value $h + (1-h)\pi/k_c$ is much larger than the minimal one, h,

$$1 + \frac{1-h}{h}\frac{\pi}{k_c} \gg 1. \tag{3.95}$$

We can see that for finite h well above zero, this is possible only for small k_c. What is more tricky, for $k_c \ll 1$ the mobility edge may emerge even in the case when the generated γ-sequence is close to a dichotomic white noise, i.e., when $1-h \ll 1$. Therefore, we should have,

$$0 < k_c \ll 1 - h \ll 1. \tag{3.96}$$

The reason of existence of the mobility edge under the conditions (3.96) is a significant contribution of the second term in the coorelator (3.94a). In spite of the fact that it has a quite small amplitude $1-h \ll 1$, this term provides extremely long-range correlations with the characteristic scale $k_c^{-1} \gg (1-h)^{-1} \gg 1$.

Now we address the function $\mathcal{S}\{K_{\gamma,h}\}(k)$ that must be non-negative in order to construct the correlated sequence γ_n with the use of the SFG method. In accordance with the definition (3.83), an appropriate analysis can be done with the following explicit expression,

$$\mathcal{S}\{K_{\gamma,h}\}(k) = 1 - \frac{\pi}{2}(1-h) + (1-h)\frac{\pi^2}{2k_c}\Theta(k_c - |k|)$$
$$+2\sum_{r=1}^{\infty}\left\{\sin\left[\frac{\pi}{2}(1-h)\frac{\sin(k_c r)}{k_c r}\right] - \frac{\pi}{2}(1-h)\frac{\sin(k_c r)}{k_c r}\right\}\cos(kr). \tag{3.97}$$

As in the previous case (3.88), the summand in the last term behaves as $\pi^3/24k_c^3 r^3$ if $r \to \infty$. Hence, at finite k_c the sum converges rapidly and *uniformly*. It is a *continuous* function of k , in particular, at $k = k_c$.

The numerical calculations of $\mathcal{S}\{K_{\gamma,h}(r)\}(k)$ performed for finite $0 < k_c < \pi$ and $0 \leqslant k \leqslant \pi$, are shown in Fig. 7. Together with Eq. (3.97) they provide us with the following empirical results.

1. Due to the last term in Eq. (3.97), the function $\mathcal{S}\{K_{\gamma,h}\}(k)$ increases with k at arbitrary values of k_c and h within both intervals $(0, k_c)$ and (k_c, π).
2. The negative jump of $\mathcal{S}\{K_{\gamma,h}\}(k)$ occurs at the same point $k = k_c$ as for the jump of the prescribed power spectrum (3.94b). The maximal and minimal values of the function are reached, respectively, at $k = k_c - 0$ and $k = k_c + 0$ for all values of the step-parameter h within $0 \leqslant h < 1$. The jump is exclusively related to the third term in Eq. (3.97). Therefore, its value reads

$$\mathcal{S}\{K_{\gamma,h}\}(k_c - 0) - \mathcal{S}\{K_{\gamma,h}\}(k_c + 0) = (1-h)\frac{\pi^2}{2k_c}. \tag{3.98}$$

3. Depending on h, the minimum $\mathcal{S}\{K_{\gamma,h}\}(k_c + 0)$ can be either negative or positive. Also, the value of $\mathcal{S}\{K_{\gamma,h}\}(k_c + 0)$ monotonically increases with an increase of h.

The presented numerical analysis can be supplemented with the following two points. First, from the treatment of the case (3.88), we know that

$$\mathcal{S}\{K_{\gamma,h}\}(k_c + 0) < 0 \qquad \text{for} \quad h = 0. \tag{3.99}$$

Second, from the definition (3.97) we can easily reveal that

$$\mathcal{S}\{K_{\gamma,h}\}(k) = 1 \qquad \text{for} \quad h = 1. \tag{3.100}$$

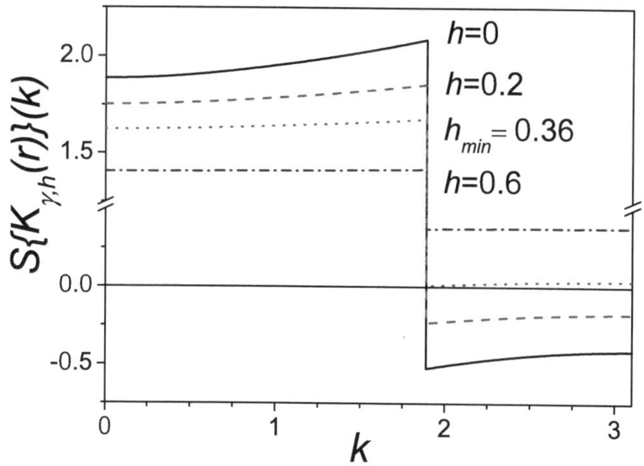

Figure 7 Function $\mathcal{S}\{K_{\gamma,h}(r)\}(k)$ vs k for few values of h at $k_c = 1.89$ (this k_c is very close to the minimum point of the curve in Fig. 6 and to the minimum point of the upper curve in Fig. 8). The function is entirely positive if $h > h_{min}$. When $h = h_{min}$, the function goes to zero solely at one point $k = k_c + 0$ and is positive otherwise. For $h < h_{min}$, the function is negative within the whole interval $k_c < k < \pi$.

Summarizing our results, we can conclude that there exists a threshold value of the step-parameter h that we refer to as h_{min}, that separates the region $0 \leqslant h < h_{min}$ in which $\mathcal{S}\{K_{\gamma,h}\}(k)$ has negative values, from the region with non-negative values,

$$\mathcal{S}\{K_{\gamma,h}\}(k) \geqslant 0 \quad \text{for} \quad h_{min} \leqslant h \leqslant 1. \tag{3.101}$$

It is clear that the threshold h_{min} obeys the equation (see Fig. 7)

$$\mathcal{S}\{K_{\gamma,h_{min}}\}(k_c + 0) = 0, \tag{3.102}$$

and depends on the correlation parameter k_c. By substitution of Eq. (3.97) into Eq. (3.102), one can rewrite it in explicit form,

$$\frac{\pi}{4}(1 - h_{min}) - \frac{1}{2} = \sum_{r=1}^{\infty} \left\{ \sin\left[\frac{\pi}{2}(1 - h_{min})\frac{\sin(k_c r)}{k_c r}\right] - \frac{\pi}{2}(1 - h_{min})\frac{\sin(k_c r)}{k_c r} \right\} \cos(k_c r). \tag{3.103}$$

The numerical solution $h_{min}(k_c)$ of this equation is displayed in Fig. 8 by the upper curve.

Eq. (3.103) can be solved analytically at small k_c, resulting in

$$h_{min}(k_c) = 1 - \left(\frac{96 k_c}{\pi^4}\right)^{1/3} + O(k_c^{2/3}) \quad \text{for} \quad k_c \ll 1. \tag{3.104}$$

This expression exhibits the limit $h_{min} \to 1$ for $k_c \to 0$.

Thus, in accordance with the study performed above, the dichotomic sequence γ_n with long-range correlator and step-wise power spectrum (3.94) can be constructed by the SFG

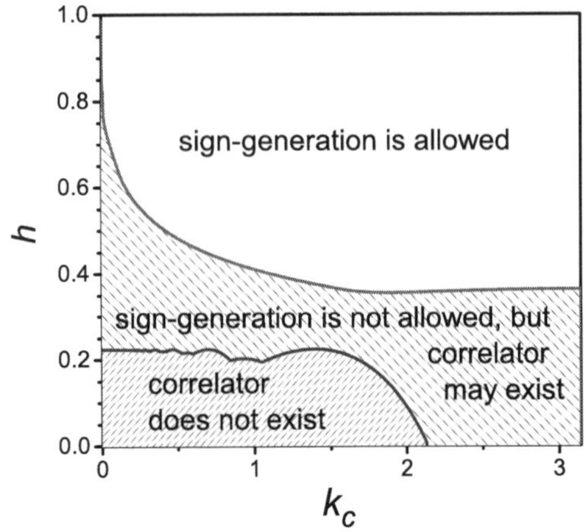

Figure 8 Space of control parameters k_c and h. Upper curve is the dependence $h_{min}(k_c)$ while low curve depicts $h_0(k_c)$. Within the lowest area in which $0 \leqslant h < h_0$, a dichotomic sequence γ_n with the correlator $K_{\gamma,h}(r)$ does not exist. The area with $h_{min} \leqslant h \leqslant 1$ allows for a creation of γ_n with $K_{\gamma,h}(r)$ by the discussed method. In the intermediate region $h_0 \leqslant h < h_{min}$ the SFG method does not work, and an existence of a dichotomic sequence with the step-wise spectrum remains an open problem.

method only if its parameters k_c and h are placed onto or above the upper line in Fig. 8. Only in this area of the (k_c, h)-plane the condition (3.101) holds true. Unfortunately, practically within this whole area the parameters k_c and h have values of the order of one, and, therefore, we cannot satisfy a quite strong requirement (3.95) in order to clearly observe a mobility edge. The only exception is a narrow vicinity of the point ($k_c = 0$, $h = 1$). Remarkably, in this vicinity due to specific dependence (3.104) of $h_{min}(k_c)$, the conditions (3.96) can be satisfied,

$$0 < k_c \ll 1 - h \leqslant 1 - h_{min}(k_c) \approx \left(\frac{96k_c}{\pi^4}\right)^{1/3} \ll 1, \tag{3.105}$$

and, consequently, the mobility edge can be achieved. So, Eq. (3.105) gives us the only (perhaps, just purely theoretical) possibility to arrange a mobility edge in the transport through the γ-sequence constructed by the SFG method.

Now we analyze the consequences of the necessary conditions formulated in Subsection 2.3.1. For the long-range correlator (3.94a) the inequalities (2.187) lead to the following restriction with respect to the step-parameter h,

$$1 - \max{}^{-1}\left\{R(r, r')\right\} \leqslant h \tag{3.106}$$

at $r \neq 0$, $r' \neq 0$ and $r \neq r'$. Here we have introduced the function

$$R(r, r') = \left|\frac{\sin k_c r'}{k_c r'} \pm \frac{\sin k_c (r - r')}{k_c (r - r')}\right| \mp \frac{\sin k_c r}{k_c r}. \tag{3.107}$$

Since by the definition $h \geqslant 0$, the requirement (3.106) is meaningful only if its l.h.s. is positive. Otherwise, it is satisfied automatically. The combination of Eq. (3.106) with the assumption $0 \leqslant h \leqslant 1$ gives rise to the relation,

$$h_0 \leqslant h \leqslant 1, \tag{3.108}$$

where new characteristic quantity h_0 is introduced,

$$h_0 = 1 - \max{}^{-1}\left\{1, \max\{R(r, r')\}\right\}. \tag{3.109}$$

This function $h_0(k_c)$ is shown in Fig. 8 by low curve. Its piecewise shape is caused by the fact that different r and r' contribute to h_0 within different intervals of k_c. Finally, when k_c becomes equal or larger than $k^* \simeq 2.139\ldots$ [see text after Eq. (3.93)], we have $h_0 = 0$ and the necessary conditions (3.108) reduce to the initial ones, $0 \leqslant h \leqslant 1$.

Thus, taking into account that the required area of the parameters k_c and h is determined by the relation, $0 < k_c < \pi$, $0 \leqslant h \leqslant 1$ of the (k_c, h)-plane, we can summarize the following. A dichotomic sequence with long-range correlator and step-wise power spectrum (3.94) does not exist within the lowest region $0 \leqslant h < h_0(k_c)$. When $h_0(k_c) \leqslant h < h_{min}(k_c)$, the dichotomic chain cannot be created by the SFG method and there is no answer whether it can be created by any other method. Finally, within the highest zone with $h_{min}(k_c) \leqslant h \leqslant 1$, one can construct desired dichotomic sequences with the use of the discussed method.

Predefined intermediate spectrum. With the SFG method, we first generate intermediate Gaussian sequence β_n and after, the dichotomic sequence γ_n. Above, we specified the correlator $K_\gamma(r)$ of a final dichotomic γ-sequence and analyzed the spectrum $\hat{K}_\beta(k) = \mathcal{S}\{K_\gamma\}(k)$ of the intermediate Gaussian β-chain, keeping in mind that the latter must be non-negative, see Eqs. (3.82), (3.85).

Below, we ask question about the type of the spectrum of γ_n, that emerges if the intermediate Gaussian sequence β_n is assumed to have given pair correlator with step-wise power spectrum of the following form,

$$K_\beta(r) = \frac{\sin(k_c r)}{k_c r}, \tag{3.110a}$$

$$\hat{K}_\beta(k) = \frac{\pi}{k_c}\Theta(k_c - |k|), \quad 0 < k_c \leqslant \pi, \ |k| \leqslant \pi. \tag{3.110b}$$

In accordance with the relation (3.80) and Fourier transforms (2.122), the corresponding correlator and power spectrum of the dichotomic γ-sequence read,

$$K_\gamma(r) = \frac{2}{\pi}\arcsin\left[\frac{\sin(k_c r)}{k_c r}\right], \tag{3.111a}$$

$$\begin{aligned}\hat{K}_\gamma(k) &= \frac{2}{\pi}\sum_{r=-\infty}^{\infty}\arcsin\left[\frac{\sin(k_c r)}{k_c r}\right]\exp(-ikr)\\ &= \left(1 - \frac{2}{\pi}\right) + \frac{2}{k_c}\Theta(k_c - |k|)\\ &\quad + \frac{4}{\pi}\sum_{r=1}^{\infty}\left\{\arcsin\left[\frac{\sin(k_c r)}{k_c r}\right] - \frac{\sin(k_c r)}{k_c r}\right\}\cos(kr).\end{aligned} \tag{3.111b}$$

Here, last representation for the spectrum $\hat{K}_\gamma(k)$ is similar to that we have employed for the function $\mathcal{S}\{\cdot\}(k)$ in its analysis [compare with Eqs. (3.83) and (3.89)]. It is noteworthy to emphasize that the summand in the last term of this representation behaves as $2/3\pi k_c^3 r^3$

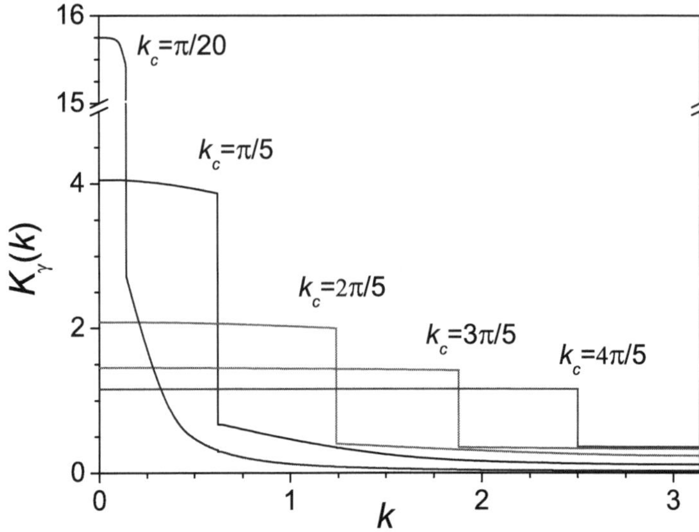

Figure 9 Power spectrum $\hat{K}_\gamma(k)$ vs k for different values of the correlation parameter k_c.

when $r \to \infty$. Hence, as above, the sum converges rapidly and *uniformly*. Therefore, it can be easily calculated numerically.

Fig. 9 presents the behavior of the power spectrum (3.111b) for dichotomic γ_n in the case of the predefined step-wise profile (3.110b) for the spectrum of intermediate sequence β_n. Since $\hat{K}_\gamma(k)$ is an even function of the wave number k, the presentation is sufficient within $0 \leqslant k \leqslant \pi$. Assuming $k_c < \pi$, from Eq. (3.111b) and Fig. 9 we can conclude:

1. The function (3.111b) is positive within the whole interval $|k| \leqslant \pi$. Therefore, it truly serves as a power spectrum and its inverse Fourier transform (3.111a) is valid correlator of the generated sequence γ_n;
2. Due to the second and last terms in Eq. (3.111b) the spectrum $\hat{K}_\gamma(k)$ decreases with an increase of k for all k_c within whole interval $(0, \pi)$;
3. The negative jump of $\hat{K}_\gamma(k)$ occurs at $k = k_c$, at the same point as for the jump of prescribed intermediate power spectrum $\hat{K}_\beta(k)$. The maximal and minimal values of $\hat{K}_\gamma(k)$ at the jump are defined at $k = k_c - 0$ and $k = k_c + 0$, respectively. The jump is exclusively formed by the second term in Eq. (3.111b). Therefore, its value reads

$$\hat{K}_\gamma(k_c - 0) - \hat{K}_\gamma(k_c + 0) = 2/k_c; \tag{3.112}$$

4. The value $\hat{K}_\gamma(k = k_c - 0)$ to the left from the jump is always positive. The value $\hat{K}_\gamma(k = k_c + 0)$ to its right is also positive for all k_c. Their ratio $\hat{K}_\gamma(k_c - 0)/\hat{K}_\gamma(k_c + 0)$ being of the order of one for finite k_c, seems to slightly increase with a decrease of k_c. However, it is saturated if the inverse correlation length k_c vanishes. Indeed, from Eq. (3.111b) we can easily get,

$$\frac{\hat{K}_\gamma(k_c-0)}{\hat{K}_\gamma(k_c+0)} = \frac{\pi + I(k_c) + (\pi/2-1)k_c}{I(k_c) + (\pi/2-1)k_c} \tag{3.113a}$$

$$\rightarrow \frac{\pi + I(0)}{I(0)} = 6.0096\ldots \text{ if } k_c \rightarrow 0. \tag{3.113b}$$

Here we have introduced

$$I(k_c) = 2k_c \sum_{r=1}^{\infty} \left\{ \arcsin\left[\frac{\sin(k_c r)}{k_c r}\right] - \frac{\sin(k_c r)}{k_c r} \right\} \cos(k_c r); \tag{3.114a}$$

$$I(0) = 2\int_0^{\infty} dx \left\{ \arcsin\left[\frac{\sin x}{x}\right] - \frac{\sin x}{x} \right\} \cos x. \tag{3.114b}$$

In spite of the divergence of the jump (3.112), the convergence of the ratio (3.113) at $k_c \rightarrow 0$ occurs because the values of the spectrum $\hat{K}_\gamma(k)$ to the left, $\hat{K}_\gamma(k_c-0)$, and to the right, $\hat{K}_\gamma(k_c+0)$, from the jump, increase with decreasing of k_c exactly in the same manner, $\hat{K}_\gamma(k_c-0) \approx 2[\pi + I(0)]/\pi k_c$ and $\hat{K}_\gamma(k_c+0) \approx 2I(0)/\pi k_c$.

In addition, it can be analytically shown that in the limit $k_c \rightarrow 0$, the correlator $K_\gamma(r)$ tends to unity, while the spectrum $\hat{K}_\gamma(k)$ turns into the Dirac delta-function,

$$\lim_{k_c \rightarrow 0} K_\gamma(r) = \frac{2}{\pi} \lim_{k_c \rightarrow 0} \arcsin\left[\frac{\sin(k_c r)}{k_c r}\right] = 1\,; \tag{3.115a}$$

$$\lim_{k_c \rightarrow 0} \hat{K}_\gamma(k) = 2\pi\delta(k). \tag{3.115b}$$

Therefore, as the correlation parameter k_c vanishes, the final dichotomic γ-sequence becomes to have extremely long-range correlations.

On the contrary, for $k_c = \pi$ the correlator $K_\gamma(r)$ reduces to the Kronecker delta-symbol, whereas the power spectrum $\hat{K}_\gamma(k)$ degenerates into unity,

$$K_\gamma(r) = \frac{2}{\pi} \arcsin\left[\delta_{r,0}\right] = \delta_{r,0}\,; \tag{3.116a}$$

$$\hat{K}_\gamma(k) = 1 \quad \text{for} \quad k_c = \pi. \tag{3.116b}$$

Thus, the final dichotomic γ_n reduces to the white-noise chain.

In summary, in spite of the fact that the jump ratio $\hat{K}_\gamma(k_c-0)/\hat{K}_\gamma(k_c+0)$ in the power spectrum $\hat{K}_\gamma(k)$ is of the order of unity for all $k_c < \pi$, we hope that the mobility edge can be observed at small enough correlation parameter $k_c \ll 1$, due to specific delta-function behavior of $\hat{K}_\gamma(k)$ itself [see Fig. 9 and Eq. (3.115b)].

3.4.4 Power correlators

Power correlator for dichotomic sequence. Here we consider an important problem of constructing a sequence with the power correlation function and corresponding spectrum,

$$K_{\gamma,p}(r) = \delta_{r,0} + (k_c|r|)^{-p}(1-\delta_{r,0}), \qquad (3.117a)$$
$$\hat{K}_{\gamma,p}(k) = 1 + k_c^{-p}\{\mathrm{Li}_p[\exp(ik)] + \mathrm{Li}_p[\exp(-ik)]\}, \qquad (3.117b)$$

$$p > 0, \quad k_c \geqslant 1, \quad |k| \leqslant \pi.$$

Here p and k_c are positive real numbers characterizing how fast the correlator decreases. Note that the parameter k_c cannot be less than one since $K_{\gamma,p}(r) \leqslant 1$. The Fourier transform $\hat{K}_{\gamma,p}(k)$ of this correlator is expressed via the polylogarithm function $\mathrm{Li}_q(z)$ that is defined by

$$\mathrm{Li}_q(z) = \sum_{r=1}^{\infty} \frac{z^r}{r^q}. \qquad (3.118)$$

For $K_{\gamma,p}(r)$ to be the correlator of a stochastic process, it is necessary to have $\hat{K}_{\gamma,p}(k) \geqslant 0$ for all k. This condition is satisfied if and only if the following inequality is fulfilled,

$$k_c \geqslant [-2\mathrm{Li}_p(-1)]^{1/p}. \qquad (3.119)$$

This result is due to the shape of spectrum $\hat{K}_{\gamma,p}(k)$ that monotonously decreases with an increase of k within the interval $(0,\pi)$, and reaches its minimal value at $k=\pi$. The r.h.s. of the condition (3.119) can be calculated in limit cases,

$$[-2\mathrm{Li}_p(-1)]^{1/p} = \begin{cases} \pi/2 - cp\,, & p \ll 1, \\ 1 + p^{-1}\ln 2, & p \gg 1\,, \end{cases} \qquad (3.120)$$
$$c = 1.029\ldots.$$

In Fig. 10 the area where power spectrum (3.117b) is non-negative, is located above the dotted lowest curve.

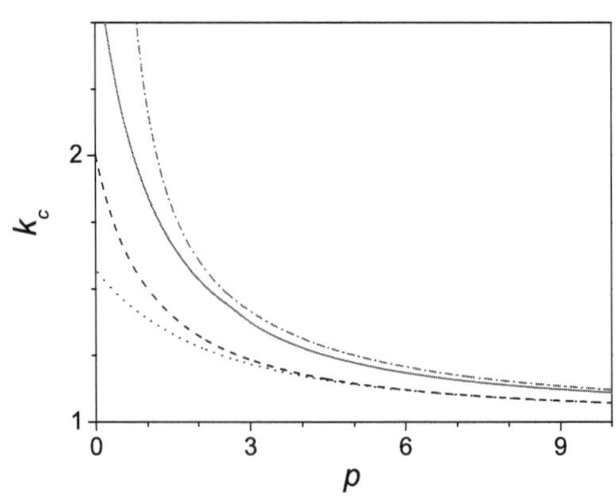

Figure 10 Various borders of the parameters p and k_c above which the following relations are fulfilled: (a) Eq. (3.119) (dotted curve), (b) Eq. (3.121)) (solid curve), (c) Eq. (3.127)) (dashed curve), (d) Eq. (3.131) (dash-dotted curve).

If $K_{\gamma,p}(r)$ is the correlator of a *dichotomic* random sequence generated by the SFG method, then $\mathcal{S}\{K_{\gamma,p}\}(k) \geqslant 0$ for all values of k. Since this function, as well as $\hat{K}_{\gamma,p}(k)$, monotonously decreases, the only condition is required,

$$\mathcal{S}\{K_{\gamma,p}\}(\pi)\geqslant 0. \tag{3.121}$$

Expanding in Eq. (3.83a) the sin-function into series, we get useful expression

$$\begin{aligned}\mathcal{S}\{K_{\gamma,p}\}(k) &= 1+\sum_{l=0}^{\infty}\frac{\pi^{2l+1}}{2^{2l+1}(2l+1)!}k_c^{-p(2l+1)}\\ &\times\{\mathrm{Li}_{p(2l+1)}[\exp(ik)]+\mathrm{Li}_{p(2l+1)}[\exp(-ik)]\}.\end{aligned} \tag{3.122}$$

For $p\to\infty$ it can be approximately calculated as

$$\mathcal{S}\{K_{\gamma,p}\}(k)\simeq 1+2\sin(\pi k_c^{-p}/2)\cos k. \tag{3.123}$$

Therefore, taking into account that $k_c\geqslant 1$, we obtain the following asymptotic of the condition (3.121)

$$k_c\geqslant 1+p^{-1}\ln 3 \qquad \text{for} \quad p\gg 1. \tag{3.124}$$

The area where the discussed SFG method is applicable located in Fig. 10 above the solid curve.

Now we analyze the necessary condition for existence of the power correlator (3.117a) for dichotomic sequence regardless of the generation method. For the sake of simplicity we consider the case $\overline{\gamma}=0$. From the inequalities (2.187) one can obtain the following relation,

$$\begin{aligned}&\max\{||r'|^{-p}\pm|r-r'|^{-p}|\mp|r|^{-p}\}\leqslant k_c^p,\\ &r\neq 0,\quad r'\neq 0,\quad r\neq r'.\end{aligned} \tag{3.125}$$

It is easy to see that the maximum with respect to r', occurs at $r'=\pm 1$ or $r'=r\pm 1$. Therefore, Eq. (3.125) can be rewritten as

$$\max_{r>0}\{1\pm(r+1)^{-p}\mp r^{-p}\}\leqslant k_c^p. \tag{3.126}$$

The last condition is equivalent to

$$k_c\geqslant(2-2^{-p})^{1/p}=\begin{cases}2-2p\ln^2 2, & p\ll 1,\\ 1+p^{-1}\ln 2, & p\gg 1.\end{cases} \tag{3.127}$$

In Fig. 10 the border corresponding to this condition is designated by dashed curve. It should be noted that Eq. (3.127) is the necessary condition, thus, if it is met, it is still not clear whether the correlator with such values of parameters p and k_c exists.

Predefined intermediate power correlator. We have found that the dichotomic sequence γ_n with the power correlator (3.117a) can be constructed by the SFG method for some values of parameters p and k_c that meet the condition (3.121). Now let us take the intermediate Gaussian sequence β_n prescribed to have the power correlator and corresponding spectrum

$$K_{\beta,p}(r) = \delta_{r,0}+(k_c|r|)^{-p}(1-\delta_{r,0}), \tag{3.128a}$$

$$\hat{K}_{\beta,p}(k) = 1+k_c^{-p}\{\mathrm{Li}_p[\exp(ik)]+\mathrm{Li}_p[\exp(-ik)]\}, \tag{3.128b}$$

$$p>0,\quad k_c\geqslant 1,\quad |k|\leqslant\pi.$$

Evidently, the condition (3.119) is implied to be met. In accordance with the relation (3.80) and Fourier transforms (2.122), the correlator and power spectrum of the generated dichotomic γ-sequence are described as

$$K_{\gamma,p}(r) = \delta_{r,0} + \frac{2}{\pi}\arcsin[(k_c|r|)^{-p}](1-\delta_{r,0}), \tag{3.129a}$$

$$\hat{K}_{\gamma,p}(k) = 1 + \frac{4}{\pi}\sum_{r=1}^{\infty}\arcsin[(k_c r)^{-p}]\cos(kr). \tag{3.129b}$$

We can assert that since Eq. (3.119) is satisfied, i.e. the spectrum (3.128b) of the intermediate β-sequence is non-negative, then the spectrum (3.129b) of the generated dichotomic γ_n is also non-negative.

When $|r| \to \infty$, the correlator (3.129a) tends to zero in accordance with the following asymptotic

$$K_{\gamma,p}(r) \simeq (k_c'|r|)^{-p}, \qquad k_c' = k_c(\pi/2)^{1/p}. \tag{3.130}$$

As was shown, the allowed values of k_c are expressed by Eq. (3.119). Therefore, the scaling parameter k_c' should satisfy the condition

$$k_c' \geqslant [-\pi \mathrm{Li}_p(-1)]^{1/p}. \tag{3.131}$$

This condition for possible values of k_c' and p is met in the area above the dash-dotted curve in Fig. 10.

Thus, the mapping of the Gaussian sequence with the power correlation function (3.128a) into the binary sequence result in the same power for the decrease of the final correlator (3.129a) expressed by Eq. (3.130). However, this behavior of the final correlator occurs only asymptotically, for sufficiently large values of $|r|$.

3.4.5 Conclusion

First, we would like to emphasize the following point that was briefly mentioned in the beginning. Our study of the correlation properties of a random dichotomic sequence γ_n is based on the example (2.169) in which two elements are "-1" and "1". On the other hand, there is a simple correspondence between this chain and a dichotomic sequence $\varepsilon(n)$ consisting of two arbitrary symbols ε_0 and ε_1,

$$\varepsilon(n) = \{\varepsilon_0, \varepsilon_1\}, \qquad n \in \mathbf{Z} = \ldots, -2, -1, 0, 1, 2, \ldots \tag{3.132}$$

The correspondence is expressed by the linear relationship,

$$\varepsilon(n) = \frac{\varepsilon_0 + \varepsilon_1}{2} \mp \frac{\varepsilon_0 - \varepsilon_1}{2}\gamma_n\,. \tag{3.133}$$

The choice of the sign is not important. It determines only into which of the symbols, ε_0 or ε_1, the initial values "-1" and "1" are converted.

In accordance with Eq. (3.133) and due to specific properties (3.77), (2.171) of the γ-sequence, the connection between the mean values and variances is as follows,

$$\varepsilon^2(n) = \frac{\varepsilon_0^2+\varepsilon_1^2}{2} \mp \frac{\varepsilon_0^2-\varepsilon_1^2}{2}\gamma_n\,; \tag{3.134a}$$

$$\overline{\varepsilon} = \frac{\varepsilon_0+\varepsilon_1}{2} \mp \frac{\varepsilon_0-\varepsilon_1}{2}\overline{\gamma}\,; \tag{3.134b}$$

$$C_\varepsilon(0) \equiv \overline{\varepsilon^2(n)} - \overline{\varepsilon}^2 = \frac{(\varepsilon_0-\varepsilon_1)^2}{4}C_\gamma(0). \tag{3.134c}$$

Analogously, the two-point correlation function $C_\varepsilon(r)$ of the ε-chain is associated with the binary correlation function $C_\gamma(r)$ of the sequence γ_n as follows

$$C_\varepsilon(r) \equiv \overline{\varepsilon(n)\varepsilon(n+r)} - \overline{\varepsilon}^2 = \frac{(\varepsilon_0-\varepsilon_1)^2}{4}C_\gamma(r). \tag{3.135}$$

The comparison of Eqs. (3.134c) and (3.135) makes obvious the equality between the normalized correlators $K_\varepsilon(r)$ and $K_\gamma(r)$,

$$K_\varepsilon(r) \equiv C_\varepsilon(r)/C_\varepsilon(0) = C_\gamma(r)/C_\gamma(0) \equiv K_\gamma(r). \tag{3.136}$$

Thus, our analysis is valid for any dichotomic sequence.

Our results can be summarized as follows. We have shown that the statistical properties of random dichotomic sequences are principally different from those known for sequences with a continuous distribution of their elements. We were able to find analytically the conditions (2.187) that can be used to know whether a binary sequence can have the desired pair correlator. Note that only these two conditions are necessary.

Another important restriction is due to the inequality (3.86) derived under quite general assumptions. We have shown that even in the well known case of an exponential decay of correlations, there are no binary sequences that can be created with the SFG method, unless the decay is sufficiently strong. This fact is very important in view of many applications.

Our specific interest was in the possibility to create, with the considered method, the binary sequences with long-range correlations described by Eqs. (3.88a) and (3.88b). We have found analytically that the function (3.88a) cannot be a pair correlator of any binary sequence. We have also examined a more general correlation function [see Eq. (3.94)] that corresponds to the generalization of the step-wise power spectrum. Our extensive examination of the signum-function method, applied to this correlation function, has revealed the regions of parameters k_c and h for which the pair correlator can emerge in a binary sequence. Correspondingly, we identified the regions where such a pair correlator cannot appear in binary sequences. For other values of the control parameters we cannot give definite answer, therefore, a further study is needed.

Finally, we analyzed an important case of the power decay of the pair correlator. Recently, the problem of the generation of random processes with power correlations has attracted much attention in the literature. Analyzing such correlators, we have found that the SFG method in principle allows us to construct binary sequences with these correlators, however, with some restrictions on the values of parameters in Eq. (3.117).

3.5 OTHER METHODS

Here we describe briefly some of known procedures enabling us to construct a correlated sequence with correlation function of concrete form, which is determined by the method of generation.

3.5.1 Fractional Brownian motion

A fractional Brownian motion [81, 9] is the integral of fractional Gaussian noise. Let t be the time, $-\infty < t < \infty$, and w be a value of the random function (w belongs to a sample space Ω). The Brownian motion $B(t, w)$ (or Bachelier, Wiener and Levy) is a random function with independent Gaussian increments such that $B(t_2, w) - B(t_1, w)$ has mean zero and variance $|t_2 - t_1|$, and such that $B(t_2, w) - B(t_1, w)$ is independent of $B(t_4, w) - B(t_3, w)$ if the intervals (t_1, t_2) and (t_3, t_4) do not overlap.

Let H be such that $0 < H < 1$ and let b_0 be an arbitrary real number. Reduced fractional Brownian motion $B_H(t, w)$ with parameter H and starting value b_0 for time $t > 0$ is defined by

$$B_H(0, w) = b_0, \tag{3.137}$$

$$B_H(t, w) - B_H(0, w) = \frac{1}{\Gamma(H + \frac{1}{2})} \left\{ \int_{-\infty}^{0} [(t-s)^{H-1/2} - (-s)^{H-1/2}] dB(s, w) + \int_0^t (t-s)^{H-1/2} dB(s, w) \right\},$$

and, similarly, does for $t < 0$. The fractional Brownian motion possesses a self-similarity properties [81].

3.5.2 Successive random additions algorithm

The random addition algorithm introduced by Voss [28] generates the 1D fractional Brownian motion. The algorithm is easily extended to higher dimensions and can produce surfaces with coastlines that are self similar fractals with a fractal dimension given by $D = 2 - H$ where H is the Hurst exponent. This variable is related to the power spectral density exponent of the process.

In the successive random addition algorithm, all points are treated equivalently at each stage of the iteration, hence the resolution at the next stage can change by any factor $r_l < 1$. A fractal is called lacunar if it contains large intervals or gaps in the fractal set (or measure). Varying the lacunarity does not alter the fractal dimension.

For simulating a rain field in two dimensions using the successive random algorithms technique, given a sample of N_n points at stage n with resolution λ, stage $n + 1$ with resolution $r_l\lambda$ is determined first by interpolating the values at the new resolution from the old N_n values. The number of points plotted at the new resolution is

$$N_{n+1} = \left(\frac{1}{r_l^{n+1}} + 1 \right)^2. \tag{3.138}$$

A random element Δ_n is then added to all N_{n+1} points. Δ_n is a Gaussian random variable with zero mean, and at stage n, with scaling ratio $l_r < 1$, the variance is given by plotted at the new resolution is

$$\Delta_n^2 \propto (r_l^n)^{2H}. \tag{3.139}$$

3.5.3 Linear Ornstein-Uhlenbeck process

The physical meaning of the Ornstein–Uhlenbeck process [22] is the behavior of the velocity of a Brownian particle under the influence of friction and immersed in a thermal bath. It is a Gaussian Markovian noise, which obeys a linear Langevin equation,

$$\dot{\eta} = -\frac{\eta}{\tau} + \frac{\xi(t)}{\tau}. \tag{3.140}$$

The process $\eta(t)$ contains a temporal memory. Parameter τ defines the characteristic time memory. The random variable $\xi(t)$ is a Gaussian white of zero mean and delta correlation, $\xi(t)\xi(t') = 2\epsilon\delta(t-t')$.

The correlation of this process is given by

$$\mathcal{K}(|t-t'|) = \langle \eta(t)\eta(t') \rangle = \frac{\epsilon}{\tau} e^{-\frac{|t-t'|}{\tau}}, \tag{3.141}$$

where ϵ is the intensity of random noise. The Fourier transform of this correlation function is

$$\hat{\mathcal{K}}(k) = \frac{2\epsilon}{1+(\tau k)^2}. \tag{3.142}$$

So, this process possesses solely the power spectrum, which decreases as k^{-2} if $k \to \infty$.

3.5.4 Successive random additions algorithm

The autoregressive process $\{h_n\}$ of order N (see, e.g. Reference [71]) belongs to the class of the continuous N-step Markov processes with the discrete time. It satisfies the equation

$$h_n + \sum_{l=1}^{N} a_l h_{n-l} = u_n, \tag{3.143}$$

where u_n is a random uncorrelated process of variance σ_u^2 and the a_l are the constants. Without loss of generality, we can assign a zero to the mean value of u_n. Then the mean of h_n is also zero. The conditional probability to have some value h_n is

$$\rho(h_n | h_{n-N}, \ldots, h_{n-2}, h_{n-1}) = \rho_U \left(u_n - \sum_{l=1}^{N} a_l h_{n-l} \right), \tag{3.144}$$

where ρ_U is the distribution function of random variable u.

The correlation function of this process is given as the linear combinations of exponential functions v_l^r $(l = 1, \ldots, N)$,

$$\mathcal{K}(j) = \sum_{l=1}^{N} A_l v_l^{|j|}, \tag{3.145}$$

where the v_l are roots of the equation

$$v^N + \sum_{l=1}^{N} a_l v^{N-l} = 0. \tag{3.146}$$

The coefficients A_l of the linear combination should be chosen according to the required form of the correlation function. The power spectrum (the Fourier transform of correlation function) of the autoregressive process is given by

$$\hat{\mathcal{K}}(k_j) = \frac{1}{2\pi}\frac{\sigma_u^2}{\sigma_h^2}\frac{1}{|1+\sum_{l=1}^{N} a_l e^{ilk_j}|^2}, \tag{3.147}$$

where σ_u is root mean square value of the u_n and σ_h is root mean square value for the h_n. The variable $k_j = 2\pi j/L$ is the spectral index, if L is the length of sequence.

Using Eq. (3.146) or (3.147) we can find the parameters a_l via the prescribed correlation function.

3.5.5 Expansion modification method

We can use this method to construct a long-range correlated binary sequence using two parameters only. The Li algorithm [29] consists in successive transitions from one finite sequence to another. Namely, a binary sequence $\{x_i^t\}$ of definite length L transforms to sequence $\{x_i^{t+1}\}$ of length $L' > L$ by the next rule:

- Each symbol 0 transforms to pair 00 (expansion) with probability $1-p$, and transforms to symbol 1 (modification) with probability p.
- Similarly, $1 \to 11$ with probability $1-q$ and $1 \to 0$ with probability q.
- Initial sequence $\{x_i^0\}$ can be chosen arbitrary. Specifically, it can consist of one symbol $x_1^0 = 0$.

Using different values of parameters p and q, the various range of the correlation functions can be obtained. Particularly, the random sequence with the power spectrum of the $1/f$ type can be generated.

4. Applications

This Section contains various physical, biological and linguistic applications of the developed theoretical tool of additive Markov chains. We begin with the physical example of the Ising spin chain with long-range interaction. The next system under consideration is a set of superconducting layers. The propagation and reflection of the Josephson plasma waves in such systems are viewed in terms of Markov chains. The designing of complex diffraction gratings with prescribed characteristics is the issue of the following Subsection. Further we study a large class of rank distributions in the stochastic systems that obey the power law. The last systems under study are the literary texts and DNA strings. Due to the wide use of the coarse graining we need to clarify some points in the Subsection 4.6. Some simple, but important examples of the dynamical systems retaining their statistical properties under discretization are given.

4.1 NON-EXTENSIVE THERMODYNAMICS

One of the basic postulates of classical statistical physics is an assumption about the particle's interaction range which is considered to be small as compared with the system size. If this condition does not hold the internal and free energies, entropy, etc. are no longer additive physical quantities. Due to this fact the definitions of temperature, entropy, etc. are not evident. The distribution function of the non-extensive system is not the Gibbs function, and the Boltzman relationship between the entropy and the statistical weight is no longer longer valid.

Non-extensive systems are common in physics [1]. An important model of the non-extensive systems is the *Ising spin chain* with elements interacting at great distances [82]. Unfortunately, the generalization of the standard Gibbsian framework for the case of arbitrary number of interacting particles is impossible if the system length is of the order of the interaction range. There exist solutions for some particular cases of the interaction of particles only (see, e.g. Reference [82]). The other results obtained for the chains with long-range interaction are either the general theorems about the existence of the phase transitions in the system or the calculations of the critical indexes [83].

Several methods for calculating the thermodynamic quantities of the long-range correlated systems have been proposed. Unfortunately, they are not well grounded and need additional justifications. One of them is *Tsallis thermodynamics* [15] based on axiomatically introduced entropy: $S^{(q)}(W) = (W^{1-q} - 1)/(1-q)$. Here W is the statistical weight and q is the parameter reflecting the non-extensiveness of the system. However, this expression does not correspond to the entropy of the Ising chain with a long but finite range of interaction. Indeed, it does not contain the size of the system and remains non-additive when the length of the system tends to infinity. Meanwhile, from a physical viewpoint, the entropy has to be additive if the system is much greater than the characteristic range of interaction.

There is a great field of research where the system length and the interaction range are macroscopically large. Thus, the problem of finding the thermodynamic quantities for the *finite* non-extensive systems is still of great importance.

A new vision on the long-range correlated systems, based on the association of the latter with *multi-step Markov chains*, has been proposed in the previous Sections. Here we assume that symbols of the binary Markov chains take on values ± 1. The unbiased *additive* Markov chain is defined by the additive *conditional probability function*,

$$P(s_i = 1 \mid T^-_{i,\infty}) = \frac{1}{2} + \sum_{r=1}^{N} F(r) s_{i-r}. \tag{4.1}$$

Here $T^{\pm}_{i,L}$ are the sets of L sequential symbols $(s_{i\pm 1}, s_{i\pm 2}, \ldots, s_{i\pm L})$. The function $F(r)$ is referred to as the memory function (see Subsection 2.2). It describes the direct interaction between the elements of the chain contrary to the binary correlation function $K(r) = \overline{s_i s_{i+r}}$ that also takes into account the indirect interactions. Here $\overline{\ldots}$ means statistical averaging over the chain. As shown in Subsection 2.2, the correlation function can be found from the recurrence relation

$$K(r) = \sum_{r'=1}^{N} 2F(r')K(r-r'), \tag{4.2}$$

that establishes the one-to-one correspondence between these two functions.

We consider the *chain of classical spins* with Hamiltonian

$$\mathrm{H} = -\sum_{\substack{j-i<N \\ i<j}} \varepsilon(j-i) s_i s_j, \tag{4.3}$$

where s_i is the spin variable taking on two values, -1 and 1, and $\varepsilon(r) > 0$ is the exchange integral of the ferromagnetic coupling. The range N of spin interaction may be arbitrary but finite.

We find thermodynamical quantities, specifically, energy and entropy, of this chain (or its parts of finite length L, comparable with interaction range N) being in thermodynamical equilibrium with a Gibbs thermostat of temperature T. The Hamiltonian (4.3) does not include the terms, corresponding to the interaction of the system with the thermostat. Nevertheless, this interaction leads to dewease in the temperature in different parts of the chain, and the thermostat fixes its temperature of the whole spin chain. A great many numerical procedures, for example, the Metropolis algorithm, were elaborated for achieving the equilibrium state. The algorithm consists of sequential trials to change the value of a randomly chosen spin. The probability of spin-flip is determined by the values of spins on *both sides* of the chain within the interaction range. Thus, the Ising spin chain in the equilibrium state can be characterized by the conditional probability, $P(s_i = s \mid T^-_{i,\infty}, T^+_{i,\infty})$, of some spin s_i to be equal to s under the condition of definite values of all other spins on *both sides* from s_i. In Subsection 2.4, a chain with the two-sided conditional probability function, which is independent of symbols at a distance of more than N, was considered. These chains were shown to be equivalent to the N-step Markov chain. So, to calculate the statistical properties of the physical object with long-range interaction, we need to find the corresponding Markov chain.

In this study, we demonstrate that the Ising spin chains with long-range interaction in a state of equilibrium are statistically equivalent to the Markov chains with some conditional probability function. Then, using the known statistical properties of the Markov chains, we calculate the thermodynamical quantities of the corresponding spin chains.

First, we present a method of defining a thermodynamical quantity Q (e.g., energy, entropy) for a subsystem of arbitrary length L (a set of L sequential particles) of any non-extensive system. This subsystem, denoted by S, interacts with the thermostat and with the rest of the system. We also denote two "border" subsystems of length N by letter B. All the other particles, except for S and B, are denoted by R:

$$\underbrace{\ldots s_{i-N}}_{R}\;\underbrace{s_{i-N+1}\ldots s_i}_{B}\;\underbrace{s_{i+1}\ldots s_{i+L}}_{S}\;\underbrace{s_{i+L+1}\ldots s_{i+L+N}}_{B}\;\underbrace{s_{i+L+N+1}\ldots}_{R}$$

The internal energy of the entire system is not equal to the sum of the internal energies of S and the rest of the system. So, the interaction with the latter makes it impossible to use the standard methods of statistical physics, e.g., the microcanonical or canonical Gibbsian ensembles.

In order to find the condition of the system equilibrium we introduce the statistical *ensemble* of the chains. We choose from it the chains with fixed $2N$ closest particles from both sides of S (with different microstate of S and R). We refer to the set of chains as *subensembles*. So, within every of 2^{2N} subensembles, the subsystem B is fixed. It plays the role of a partition wall conducting energy and keeps its internal energy unchanged. The energy of system $S+R$ equals to the sum of energies of S and R and their energy of interaction with B. Thus, within every subensemble we can use the equilibrium condition between S and R, analogous to that for extensive thermodynamics:

$$\frac{\partial \ln W_S(E_S|B)}{\partial E_S} = \frac{\partial \ln W_R(E_R|B)}{\partial E_R}, \tag{4.4}$$

where $W_S\,(E_S|B)$ and $W_R(E_R|B)$ are the statistical weights of systems, in which the energy of S is E_S, those of R is E_R, and B is fixed. We refer to these statistical weights as the *conditional statistical weights*. It should be emphasized that these weights are not totally macroscopical quantities because of their dependence on the microscopic state B.

In a similar way, we introduce any conditional thermodynamical quantity $Q(\cdot|B)$. The real quantity $Q(\cdot)$ is the conditional one, averaged over the subensembles with different environments B:

$$Q(\cdot) = \langle Q(\cdot|B)\rangle_B. \tag{4.5}$$

Using this way of calculating the thermodynamical quantities, we do not need to find the distribution function of the system S. Nevertheless, it is quite appropriate for the non-extensive systems and yields the thermodynamical values that could be measured experimentally. Generally speaking, quantity Q in Eq. (4.5) corresponds to the finite (micro-, meso- or macroscopic) system. Specifically, we can consider it in the thermodynamical limit $L \to \infty$, but we focus our attention on the finite L. If the system is in the thermal contact with the Gibbs thermostat, within every subensemble with a fixed B, the equilibrium condition between thermostat and subsystem S yields the temperature of S equal to T. Thus, the averaged temperature of the subsystem S is also T.

The conditional entropy can be introduced as the logarithm of the conditional statistical weight: $S(E_S|B) = \ln W(E_S, S|B)$. Equality $dS(E_S|B) = dE_S/T$ is fulfilled for the conditional quantities S and E. Meanwhile, such a relation is not valid for the averaged entropy and energy.

Note that the presented method for calculating the thermodynamical quantities is rather general and can be applied to other quantities, e.g., to the probability for some spin to take on the definite value under the condition of the particular environment.

Now we proceed to the application of this method for the analysis of the *Ising spin chain* with long-range interaction. The theory of Markov chains is built around the expression for the *conditional probability function*. So, in an effort to find a Markov chain that corresponds to a given Ising system, we have to find its conditional probability function. To this end, we consider one spin s_i as a system S and the conditional probability P as a quantity Q in the above-mentioned method. In the extensive thermodynamics, this probability is determined by the Gibbs distribution function. According to our method, one can find that the probability of event $s_i = 1$ under the condition of the fixed spins from B is given as

$$p(s_i = 1|B) = \left(1 + \exp\left(-\sum_{r=1}^{N} \frac{2\varepsilon(r)}{T}(s_{i-r} + s_{i+r})\right)\right)^{-1}. \tag{4.6}$$

It should be pointed out that this expression is similar to the Glauber formula [84] written in some other terms. Our result has been that the conditional probability of the s_i occurring is determined by two subsystems of length N on both sides of s_i only and does not depend on the remoter spins. As mentioned above, this is a property of the N-step Markov sequences.

To arrive at an explicit expressions for the *energy and entropy* of the subsystem S we consider the case of *weak interaction*,

$$\sum_{r=1}^{N} \varepsilon(r) \ll T. \tag{4.7}$$

It should be noted that the interaction range N can be taken as ∞, for example, when the series $\sum_r \varepsilon(r)$ converges. In this case, Eq. (4.6) corresponds approximately to the additive Markov chain with the memory function $F(r) \approx \varepsilon(r)/2T$ and, thus, its binary correlation function is

$$K(r) \approx \delta_{r,0} + \varepsilon(r)/T. \tag{4.8}$$

Figure 1 shows that the Ising spin chain being in equilibrium state is statistically equivalent to the Markov chain. The solid circles describe the binary correlation function of the additive Markov chain with the memory function $F(r) = \varepsilon(r)/2T$. The open circles correspond to the correlation function of the spin chain being equilibrated by the Metropolis scheme.

It can be shown that the Metropolis scheme also yields the same expression (4.6) for the conditional probability. Due to this fact we do not need to use the Metropolis algorithm to find an equilibrium state of the spin chain but can generate a Markov chain with a corresponding conditional probability function.

Now we proceed to examine the *binary spin chain* determined by the Hamiltonian (4.3) with $s_i = \pm 1$ and find the non-extensive energy and entropy of subsystem S containing L

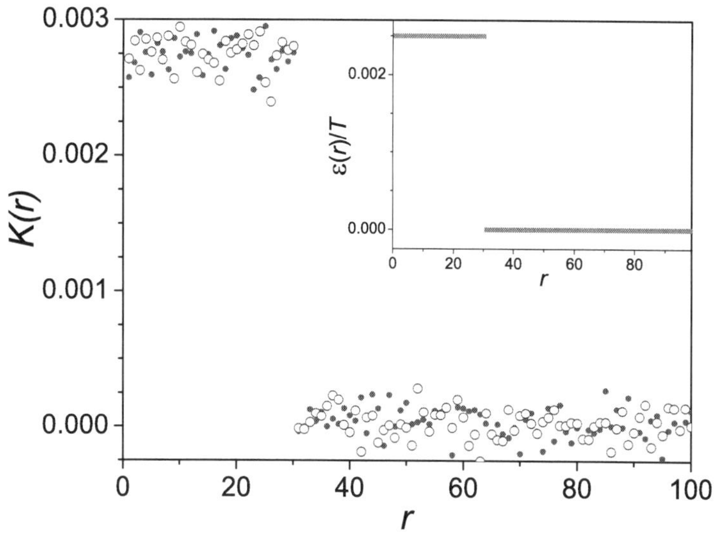

Figure 1 The correlation functions of the additive Markov chain determined by memory function $F(r) = \varepsilon(r)/2T$ (solid circles) and the spin chain being equilibrated by the Metropolis scheme (open circles). The inset represents the function $\varepsilon(r)/T$, that equals to 0.0025 for $r \leq 30$. The interaction range $N = 30$, the number of the particle in chain is 10^6, the number of attempts to invert spins in Metropolis algorithm is 10^7.

particles. Since the explicit expressions for the correlation function were derived for the additive Markov chains with the small memory function only, we suppose the energy of interaction to be small as compared to the temperature, Eq. (4.7).

The proposed method of finding thermodynamic quantities is considerably simpler while finding the *energy* of the system S of length L. The energy is the averaged sum of products of pairs of spin values,

$$E_L(T) = \overline{- \sum_{j=i}^{i+L} \sum_{k=j-N}^{j+N} \varepsilon(|k-j|) s_k s_j + \sum_{j,k=i}^{i+L} \varepsilon(|k-j|) s_k s_j}. \tag{4.9}$$

The first term accounts the interaction of each spin of the system S with $2N$ surrounding ones. The mutual interactions between the spins of S are accounted twice, thus the second term subtracts the corresponding summands. The overline represents an averaging over the whole ensemble of the chains as distinct from the $\langle \cdot \rangle$ that means averaging over the configurations of B.

Formally, Eq. (4.9) depends on the index i of the first spin in the system S. However, we study the homogeneous systems, so function $E_L(T)$ in Eq. (4.9) does not contain i as an argument. Energy in Eq. (4.9) can be calculated via the binary correlation function only with the arguments, less than N, without finding conditional energies. Using Eq. (4.8), we arrive at

$$E_L(T) = -\left(\epsilon_N^2 L + \sum_{i=1}^{N} \varepsilon^2(i) \min\{i, L\} \right) / T, \tag{4.10}$$

with

$$\epsilon_N^2 = \sum_{i=1}^{N} \varepsilon^2(i). \tag{4.11}$$

If the system length is much greater than memory length N this expression yields extensive energy. Indeed, in this case subsystems S and B interact nearly extensively. In the opposite limiting case we get:

$$E_L(T) \approx -(4\epsilon_N^2 L - \varepsilon^2(1)L^2)/2T, \quad L \ll N. \tag{4.12}$$

If we regard the subsystem S as the *whole* chain of the length L, forming the circle, the additive energy is

$$E_L(T) = -\epsilon_N^2 L/T. \tag{4.13}$$

It is obtained analogously to Eq. (4.9) using the expression for the correlation function. In this case there are no subsystems B and R though $N \neq 0$. This result is very natural because the whole chain interacts extensively with thermostat, so the extensive thermodynamics is valid.

It is seen, that the non-extensive energy is expressed in terms of binary correlation functions only. This is not correct for other thermodynamical quantities, e.g. the *entropy* of the system S. Formally, to find the entropy, we have to calculate all conditional entropies by *integration* of $dS(E|B) = dE(S|B)/T$, and then *to average* the result over all realizations of the borders. However, at high temperatures, to a first approximation in the small parameter $\sum_{r=1}^{N} \varepsilon(r)/T$ we can change the order of these operations and calculate the averaged entropy by integrating this formula taken with the averaged energy. A constant of integration is such that the chain is completely randomized at $T \to \infty$, and its entropy is equal to $\ln 2^L$. Thus, we obtain

$$S_L(T) = L \ln 2 + E_L(T)/2T. \tag{4.14}$$

Here $E_L(T)$ is determined by one of Eqs. (4.10)–(4.13). This expression describes the non-extensive entropy of the system S because of the non-linear dependence of $E_L(T)$ and $S_L(T)$ on L. However, while the energy is non-extensive in the main approximation, the entropy is non-extensive as a first approximation in the parameter $\sum_{r=1}^{N} \varepsilon(r)/T$. The dependences of the non-extensive energy and entropy on size L of the system S are given in Fig. 2 for step-wise interaction $\varepsilon(r)$.

It should be emphasized, that knowing the energy and entropy we can find some other thermodynamic quantities. For example, at high temperatures the heat capacity can be determined in a similar way as for the entropy calculating. We can use the classical formula $C_V(T) = TdS_L(T)/dT$ with averaged entropy from Eq. (4.14) and obtain the heat capacity value as $C_V = -E_L(T)/T$. This simple procedure of its finding is valid for the case of high temperatures as the main approximation only. In the general case, $C_V(T)$ is not determined by $C_V(T) = TdS_L(T)/dT$. This relation holds for the conditional quantities only.

At high temperatures, we calculated the averaged non-additive energy and entropy without using the conditional ones. In the opposite limiting case of low temperatures, the

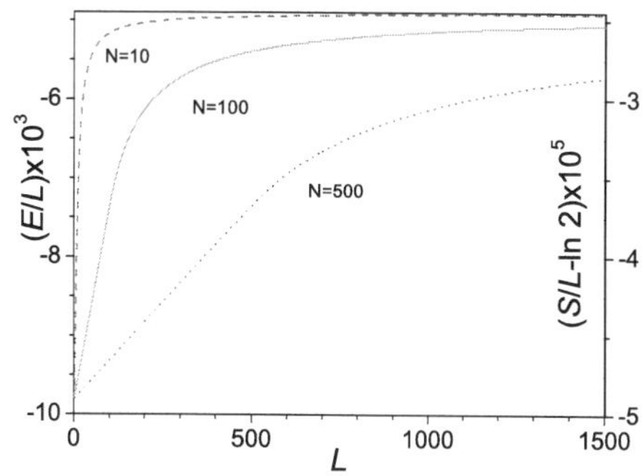

Figure 2 The specific non-extensive energy $E \times 10^3/L$ and entropy $(S/L - \ln 2) \times 10^5$ vs. the size L of system for step-wise interaction $\varepsilon(r)$. The memory length N is indicated near the curves. The constant limiting value of energy is $-4\epsilon_N^2/2T \approx -5 \times 10^{-3}$.

calculations of conditional quantities proves to be necessary. The corresponding calculations will be published.

Thus, we have suggested the method of evaluating thermodynamic characteristics of non-extensive systems. The value of certain thermodynamical quantity can be obtained by averaging the corresponding conditional quantity. This method is applied to the Ising spin chain. The explicit expressions for the non-additive energy and entropy are deduced in the limiting case of high temperatures as compared to the energy of spins interaction. At high temperatures, the equilibrium Ising chain of spin turns out to be equivalent to the additive multi-step Markov chain.

4.2 JOSEPHSON CHAINS AND SUPERCONDUCTING MULTILAYERS

In this Subsection we show that the wave propagation in both discrete Josephson transmission lines and superconducting multilayers can be mapped onto the Anderson tight-binding model originally developed for quantum particle moving through discrete 1D site potentials. When the parameters of Josephson junction randomly fluctuate, the waves cannot propagate due to the Anderson localization. For fluctuations with long-range correlations, the transparency edges can be observed. With appropriate choice of the correlation function, an arbitrary transparency frequency windows for Josephson plasma waves can be achieved, which could be useful for THz filters.

4.2.1 Introduction

The growing number of studies of Josephson structures is partly motivated by the subterahertz (up to several hundreds GHz) frequency range of the electromagnetic waves (so-called Josephson plasma waves) propagating in these systems. Controlling Josephson plasma waves (JPWs) in Josephson arrays could be potentially useful for applications in physics, astronomy, chemistry, biology, and medicine, including THz imaging, spectroscopy, tomography, medical diagnosis, health monitoring, environmental control, as well as chemical and biological identification [85]. These THz electromagnetic waves interact nonlinearly with the Josephson medium itself and with magnetic flux quanta (Josephson vortices), which, in turn, can be conveniently manipulated by varying an in-plane magnetic field and/or an out-of-plane electric current. Such an unprecedented level of controllability has been used to propose a set of well-integrated THz devices [86, 87, 88, 90, 89], including generators, tunable filters, detectors, wave mixers, lenses, converters, and amplifiers.

Furthermore, recent achievements in nanotechnology allow the fabrication of parallel-coupled one dimensional (1D) Josephson junction chains [91, 92] (Fig. 3a), known as discrete Josephson transmission lines, and artificial superconducting multilayers [93] (Fig. 3b) with prescribed properties. For instance, we can controllably introduce some degree of disorder in these nanodevices. Since the theoretical models describing the Josephson arrays are well defined, it offers the opportunity of theoretically exploring and experimentally testing several fundamental concepts of low-dimensional physics and employing these for further developing superconducting THz devices.

In this Subsection we continue to study transport properties of long-range correlated disordered media. Here we describe an analogy between the behavior of a quantum particle and the sub-terahertz wave propagation in 1D Josephson arrays and superconducting multilayers. For instance, it has already been shown [88, 94] that the frequency intervals where the electromagnetic Josephson plasma waves (JPWs) can propagate form a zone structure if a periodic superlattice of Josephson vortices is induced by an in-plane magnetic field in superconducting multilayers or a long 1D Josephson junction. This JPW photonic crystal [88], with forbidden frequency zones tuned by the in-plane magnetic field, is a much more controllable analog of both: the band structure in 1D conductors and standard photonic crystals [95]. Here, we use this analogy (see Table I) between JPWs and quantum particles to show how to manipulate the THz and sub-THz electromagnetic wave localization and the transparency of Josephson chains and artificial superconducting multilayers via a controllable change of their disorder. In particular, we predict the frequency windows of transparency for samples with long-range correlated disorder. The effective random potential can be either continuously distributed or take a discrete number of values. We show that even a dichotomic (or binary) potential (i.e., discrete potential taking randomly only two possible values) can be used to construct transmission lines with any desirable transparency window (acoustic analogs of these have been studied in Reference [96]). Moreover, for superconducting multilayers, the transparency edge can also be observed as a function of either the incident angle or frequency of the sub-terahertz radiation, which could be useful for tailoring the electromagnetic wave spectrum of the array [88]. The Josephson structures proposed here can be potentially useful for THz and sub-THz filters which are sensitive to either the frequency or the propagation direction of electromagnetic waves. The experimental realization of this proposal would be another example of controlled Anderson localization, among few, including experiments with GaAs-AlGaAs random superlattices [97] and with microwave transmission spectra through a single-mode waveguide with inserted correlated scatterers [70].

4.2.2 Josephson junction chains

Designing a potential for JPWs in a quasi-1D Josephson array We consider JPWs propagating along the discrete Josephson transmission line schematically shown in Fig. 3a. The gauge-invariant phase difference ϕ_n, determining the electric and magnetic fields at the

	Quantum particles	Josephson plasma waves
Model	Tight-binding model	Linearized sine-Gordon model
Mapping	Energy E quasi-momentum k	Frequency ω Wave number q
Randomness via	Random potential	Randomness in the Josephson junctions
Transparency edge via correlations	Yes	Yes, by changing either the incident angle (for superconductors) or the frequency (for the Josephson transmission line)
How to observe	E.g., metal-insulator transition, transmission coefficient	Transmissivity changes: from transparent to opaque for JPW

Comparison beyond undamped, linear response

Nonlinearity	Can occur due to interparticle interactions, but not easily controllable	Controllable nonlinearity due to the nonlinear J versus ϕ ($J \propto \sin\phi$)
Damping	Can occur, but not controllable	Controllable by changing temperature
Tunability	No	Yes, by locally changing currents or fields

Table 1 Comparison between localization of quantum particles and Josephson plasma waves. Here "tunability" refers to the ability to change the transport properties of a system by tuning an external parameter. This ability to tune and control transport could be useful for applications.

nth junction, obeys the discretized version of the perturbed sine-Gordon equation (see, e.g., Reference [98]):

$$\frac{1}{\omega_J^2}\frac{d^2\phi_n}{dt^2}+\frac{1}{\omega_R}\frac{d\phi_n}{dt}+\frac{J_c^{(n)}}{J_c}\sin\phi_n = \frac{1}{a^2}\left(\phi_{n+1}+\phi_{n-1}-2\phi_n\right)+j_n \tag{4.15}$$

Here, ω_J is the Josephson plasma frequency, $a=d/\lambda_J$ is the discreteness parameter, d is the spatial period of the array, λ_J is the Josephson penetration depth, ω_R is the dissipation frequency, $J_n^{(n)}/J_c$ is the ratio of the critical current $J_c^{(n)}$ in the nth junction and the average critical current J_c, and $j_n = J_n/J_c$ is the normalized external current flowing through the nth junction. In the experiment [98], the discreteness parameter can be chosen arbitrarily, e.g., it can take on any value between 0.1 and 10.

Let us first consider the case when all contacts are identical, $J_c^{(n)}/J_c = 1$, while we can change the driving current j_n through the junctions. The solution of Eq. (4.15) for plasma waves can be written in the form $\phi_n = \phi_n^{(0)} + \varphi_n$, with $\varphi_n \ll 1$, where

$$j_n = \frac{1}{a^2}\left(\phi_{n+1}^{(0)}+\phi_{n-1}^{(0)}-2\phi_n^{(0)}\right)-\sin\phi_n^{(0)} \tag{4.16}$$

and

$$\frac{1}{\omega_J^2}\frac{d^2\varphi_n}{dt^2}+\frac{1}{\omega_R}\frac{d\varphi_n}{dt}+\left(1+\delta_n\right)\varphi_n=\frac{1}{a^2}\left(\varphi_{n+1}+\varphi_{n-1}-2\varphi_n\right) \tag{4.17}$$

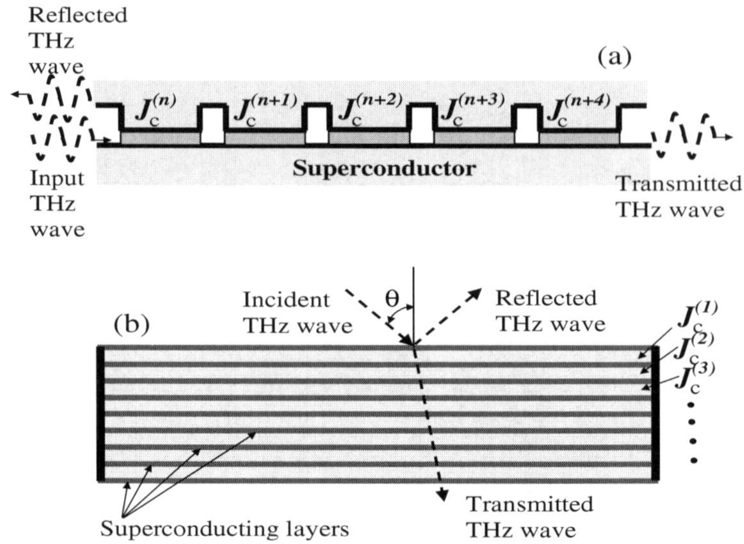

Figure 3 Schematic diagram of the parallel-coupled 1D Josephson junction array, known as discrete Josephson transmission line (a), and an artificial superconducting-nonsuperconducting multilayer (b). For both systems, disorder can be introduced by varying the Josephson critical currents $J_c^{(n)} = J_c(1+\delta_n) = J_c(1 \pm \delta)$. These disordered Josephson transmission line (a) and artificial multilayers (b) can be mapped to a binary (dichotomic) chain with symbol "1" corresponding to $+\delta$ and "0" referring to $-\delta$.

with $\delta_n = \cos\phi_n^{(0)} - 1$ which can be seen as a tunable (by changing j_n) "potential" for JPW. In this sense, Eq. (4.16) can be considered as a recipe to calculate the current distribution which has to be used to obtain a desirable set of $\phi_n^{(0)}$ and, thus, δ_n. Instead of currents j_n, magnetic fields applied between adjacent Josephson junctions can also be used to produce a tunable potential for JPWs.

An alternative way to create a potential δ_n for JPWs would be the fabrication of a discrete transmission line with a desirable distribution of the parameters. For instance, the critical current of the nth junction can vary, taking the values $J_c^{(n)} = J_c(1+\delta_n)$ with the zero-averaged δ_n, i.e., $\langle\delta_n\rangle = 0$. Considering JPWs with $\phi_n = \varphi_n \ll 1$, we return to Eq. (4.17), but now with a fixed potential.

In practice, we can never change the critical current alone. For example, we can design the junction area to be different from junction to junction. This will give an extra fluctuating coefficient in front of the first term of Eq. (4.17). Taking into account this coefficient results in the renormalization of the fluctuating parameter δ_n. The exact expression for δ_n depends on the design of the experiment.

4.2.3 Mapping the Josephson chain onto a tight-binding model

In order to map the problem of propagating JPWs in a transmission line (Fig.3a), we need to neglect both energy dissipation and nonlinearity. The dissipation term in Eq. (4.17) can be safely omitted for low enough temperatures, while the nonlinearity becomes important [90] only for high enough amplitudes of the plasma waves. As a result, the equation describing the propagation of linear JPWs in this discrete system can be rewritten as

$$\varphi_{n+1} + \varphi_{n-1} + U_n \varphi_n = E \varphi_n \tag{4.18}$$

with

$$U_n = a^2 \delta_n; \quad E(\omega) = 2 - a^2 \left(\frac{\omega^2}{\omega_J^2} - 1 \right). \tag{4.19}$$

Below we assume $U_n \ll 1$, which can be easily satisfied by an appropriate choice of δ_n and a.

Equation (4.18) is the same as in the tight-binding model proposed by Anderson [99] for the description of quantum particle motion in discrete 1D uncorrelated site potentials U_n. For $U_n = 0$ (i.e., at $\delta_n = 0$) the particle is delocalized,

$$\varphi_n = \varphi_0 \exp(iqn), \quad E = 2\cos q. \tag{4.20}$$

This corresponds to the known result [98] that JPWs with the wave number q and

$$\frac{\omega^2}{\omega_J^2} = 1 + \frac{4}{a^2} \sin^2(q/2) \tag{4.21}$$

can propagate in the transmission line. For periodic U_n (i.e., periodic δ_n), the amplitude in Eq. (4.20) slightly oscillates, $\varphi_0(n) = \varphi_0 + \phi(n); \quad |\phi(n)| \ll \varphi_0$, with the period of the potential U_n. For this case, the zone structure of E (photonic crystal for the JPW spectrum (see, Ref. [88]) or energy zone structure for quantum particles) occurs. For uncorrelated (white noise) random potentials U_n, the envelope curve $\varphi_0(n)$ decays exponentially due to localization. According to Thouless [100], the localization length l_0 for quantum particles obeys the relation,

$$l_0(E) = 8d \sin^2 q(E)/U_0^2, \tag{4.22}$$

where d is the spacing between the potential sites, $U_0^2 = \langle U_n^2 \rangle$ is the variance of the random potential. For Josephson 1D arrays, the localization length $l_0^{\mathrm{JPW}}(\omega) = l_0(E(\omega))$ can be rewritten as:

$$l_0^{\mathrm{JPW}}(\omega) = \frac{8da^2}{U_0^2} \left(\frac{\omega^2}{\omega_J^2} - 1 \right) \cdot \left[1 - \frac{a^2}{4} \left(\frac{\omega^2}{\omega_J^2} - 1 \right) \right]. \tag{4.23}$$

Using this result, we can conclude that the transparency $T(\omega)$ of the transmission line is exponentially small, i.e.,

$$T(\omega) \propto \exp\left[-\frac{L}{l_0^{\mathrm{JPW}}(\omega)} \right], \tag{4.24}$$

if the sample length L exceeds the localization length l_0^{JPW} (see, e.g., Ref. [10]). For frequencies close to ω_J, localization occurs if the number N of junctions in the array exceeds $2/U_0^{2/3}$. Here we take into account that $q > q_{\min} = 1/N$. The transmission line becomes completely nontransparent (for any frequencies) if $N \gg 8/U_0^2$.

4.2.4 Transparency edge due to correlated disorder

Correlations in the random potential introduce some degree of order and thus lead to an increase in the localization length. According to References [103, 102, 104, 105, 25], the localization length $l(E)$ for quantum particles is determined by the Fourier transform of the normalized correlator

$$\xi(r) = \frac{\langle U_n U_{n+r} \rangle}{U_0^2}, \tag{4.25}$$

where $\langle ... \rangle$ denotes the average over the chain:

$$\begin{aligned} l(E) \equiv l(E, q(E)) &= \frac{l_0(E)}{\tilde{\xi}(2q)}, \\ \tilde{\xi}(q) &= 1 + 2 \sum_{r=1}^{\infty} \xi(r) \cos(qr) \end{aligned} \tag{4.26}$$

with $q(E)$ defined by Eq. (4.20). For the white-noise potential, $\xi(r) = \delta_{0,r}$, and Eq. (4.26) gives $l(E) = l_0(E)$, i.e., we obtain the previous result Eq. (4.22). In the case of short-range correlations, the correlator $\xi(r)$ for several values of r is different from zero and the Fourier component $\tilde{\xi}(k)$ can only vanish at some discrete energies, where $\tilde{\xi}(q) = 0$. Thus, short-range correlations of random potentials can result in extended states [106] at several isolated energies.

If the correlations are long-range (e.g., $\xi(r)$ decays as a power law), a continuum of extended states can appear in the energy spectrum [96, 25, 72], since $\tilde{\xi}(q)$ can vanish in an interval of q. In other words, controllable long-range correlations facilitate the observation of the metal-insulator transition [70]. Therefore, transparency windows can be observed in the transmission line, if we fabricate either a transmission line with a locally changeable external current and apply j_n with desirable correlations, or a junction array with fixed appropriate correlations of J_c^n.

There exist several methods of generating partially-random sequences with desirable correlation functions. For example, a random set of scattering centers, with *continuously distributed parameters* [70], was used to manipulate microwave transmission spectra through a single-mode waveguide. However, this requires a precise control of the system parameters. It was shown in Subsection 2.1 (see References [31, 107] also) that random sequences with any prescribed correlation function can be constructed using additive many-step Markov chains with long-range memory function [122]. In this case, it is possible to use only *two values* for the random variable, which is much easier to control during the fabrication process. Thus, for example, we can consider arrays of junctions where the Josephson current takes only two values, $J_c^{(n)} = J_c(1 \pm \delta)$. Previous related work on acoustic wave propagation in semiconductor multilayers appears in Ref. [96].

Using our Eqs. (4.19–4.26), we now calculate the frequency dependence of the transmission coefficient

$$T/T_0 = T/T(\delta = 0) = \exp(-L/l^{\mathrm{JPW}}(\omega)), \tag{4.27}$$

normalized to the value of T at zero fluctuations. A correlated random binary sequence (the concrete realization of binary sequences in Reference [108] which were used when plotting

Figs. 4–6) of 10^5 symbols (either zero or one) was generated by means of the method described in Subsection 3.2.1. The sizes of the samples (number of symbols) were chosen to get a significant filtering effect which gradually decays for smaller samples.

The results of the calculations of $T(\omega)/T_0$ are presented for transmission lines in Fig. 4a, 5a. The dashed lower curves correspond to uncorrelated disorder, which strongly suppresses the transmissivity for long enough samples. The solid upper curves describe the transmissivity of the samples with long-range correlated disorder having the correlation function shown in Figs. 4b,c and 5b,c. The correlation function in Fig. 4 was chosen so that it suppresses the transmission of THz waves at the middle $(\omega/\omega_J) \approx 1 + 2/a^2$ of the spectrum located in

$$1 < \frac{\omega}{\omega_J} < \sqrt{1 + \frac{4}{a^2}}, \tag{4.28}$$

while the transmission of THz waves with frequencies near the edge of the spectrum is suppressed for the correlations used here to obtain Fig. 5. Thus, it is clearly seen that introducing correlations allows us to obtain transparency windows and frequency-selective filtering.

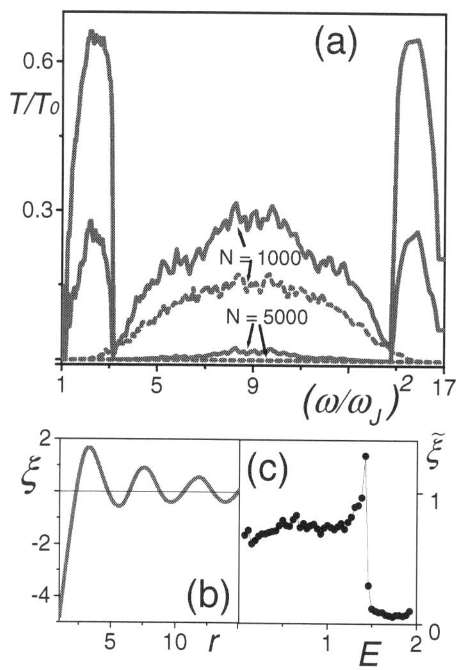

Figure 4 Controlling transparency windows in frequency-selective THz filters (Fig.3a). (a) The dimensionless transmission coefficient $T/T_0 = T/T(\delta=0) = T(\omega)/T(\omega,\ \delta=0)$ versus dimensionless frequency ω/ω_J for different lengths $N = L/d$ of Josephson chains (Fig. 3a). Blue dashed lines correspond to $T(\omega)$ for uncorrelated disorder, while the red solid curves correspond to $T(\omega)$ with disorder in the correlation functions which are shown in the (b) and (c) panels. Spatial correlations associated with dimensionless $\xi(r)$ (normalized by a), shown in panel (b), produce correlations in the frequency domain: shown in the Fourier transform $\tilde{\xi}(E)$ with dimensionless E in panel (c). The parameters used here are: $a = 0.5$; dispersion $\propto U_0^2 = 10^{-2}$. It is clearly seen that varying the correlation function ξ controls the transparency windows in $T(\omega)$.

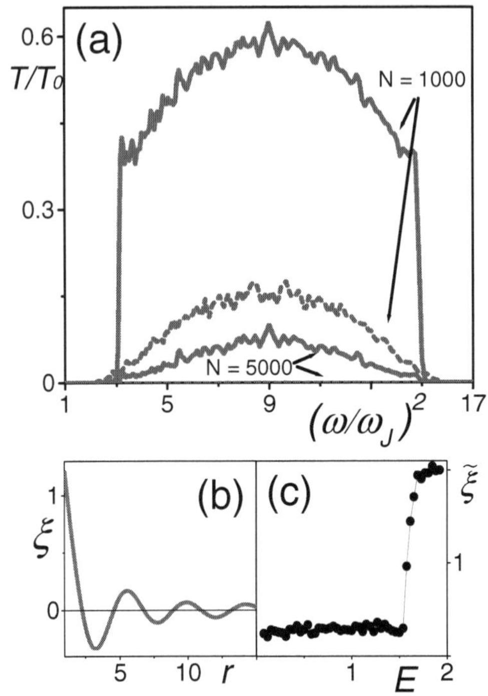

Figure 5 Controlling the transparency windows in frequency-selective THz filters (Fig. 3a): Same as in Fig. 4, but with dimensionless correlation functions ξ shown in Fig. 3(b) and (c). Comparing Figs. 4 and 5, it is clearly seen that varying the correlation function ξ controls the transparency windows in $T(\omega)$. Notice the sharp change in the correlation function $\tilde{\xi}(E)$ in both, Figs. 4(c) and 5(c).

4.2.5 The transparency edge in superconducting multilayers

Now we consider superconducting multilayers with fluctuating Josephson current, $J_c^{(n)} = J_c(1+\delta_n)$, between the nth and $(n+1)$th layers. This system is described by the coupled sine-Gordon equations [109]

$$\left(1 - \frac{s^2}{\lambda_{ab}^2}\Delta_n\right)\left\{\frac{\partial^2\varphi_n}{\partial t^2} + \omega_J^2 \sin(\varphi_n)\right\} - \frac{c^2}{\varepsilon}\frac{\partial^2\varphi}{\partial x^2} = 0, \tag{4.29}$$

where λ_{ab} is the in-plane London penetration depth, s is the interlayer spacing, ε is the interlayer dielectric constant, and the operator Δ_n is defined as $\Delta_n f = f_{n+1} - 2f_n + f_{n-1}$. Linearizing this equation ($\sin(\varphi_n) \to \varphi_n$), and considering a monochromatic THz wave with frequency ω and wave vector ω/c incident from the vacuum to the ab-plane of the sample, we obtain

$$\varphi_n(x,t) = \tilde{\varphi}_n \exp\left(i\omega t - i\frac{\omega \sin\theta}{c}t\right), \tag{4.30}$$

where θ is the incident angle of the THz waves (see Fig. 3). Here we assume that the in-plane component of the wave vector cannot change at the sample boundary. For $\tilde{\varphi}i_n$ we derive Eq. (4.29) in the same form as Eq. (4.18) with

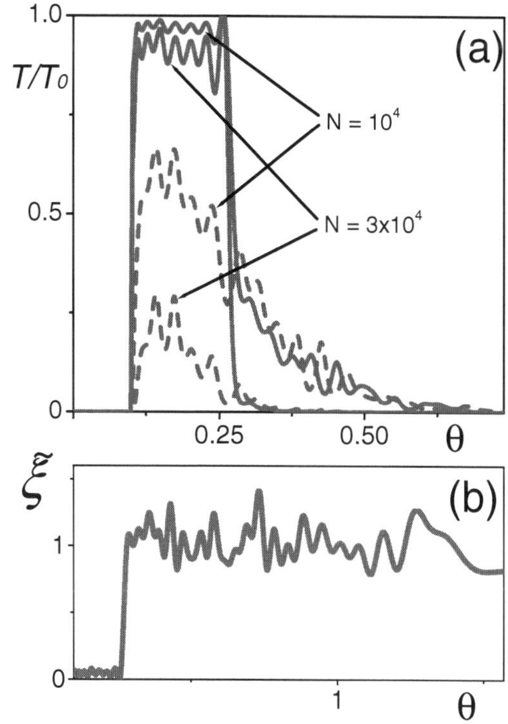

Figure 6 Controlling transparency windows in angle-selective THz filters. (a) The dimensionless transmission coefficient $T/T(\delta=0)$ versus the incident angle θ of THz radiation (see Fig. 3b) for different thickness $N=L/d$ of artificial multilayers (Fig. 3b). Blue dashed lines correspond to $T(\theta)$ for uncorrelated disorder in the Josephson critical currents between layers, while the red solid curves correspond to $T(\theta)$ with correlated disorder. Correlations are described by the dimensionless angular-correlation function $\xi(\theta)$ shown in the panel (b). The parameters used here are $\lambda_{ab}/\lambda_c=200, \delta=10^{-3}$, $(\omega^2-\omega_J^2)/\omega_J^2=10^{-3}$, and $\varepsilon=10$.

$$E=2-\frac{s^2}{\lambda_{ab}^2}\left(\frac{\omega^2\sin^2\theta}{\varepsilon(\omega^2-\omega_J^2)}-1\right), \tag{4.31}$$

$$U_n=\frac{s^2}{\lambda_{ab}^2}\frac{\omega^2\sin^2\theta}{\varepsilon(\omega^2-\omega_J^2)^2}\cdot\delta_n. \tag{4.32}$$

In order to obtain the expression (4.32) for random potentials, we use the relation

$$\left(\omega_J^{(n)}\right)^2=\frac{8\pi esJ_c^{(n)}}{\hbar\epsilon}=\omega_J^2(1+\delta_n) \tag{4.33}$$

with $\omega_J^2=8\pi esJ_c/\hbar\epsilon$.

Thus, we show that the coupled sine-Gordon model can be also mapped to the tight-binding model (4.18). For samples with correlated disorder, one can observe the transmissivity window with respect to either the incident angle θ or frequency ω offering the possibility of filtering sub-THz and THz radiation with respect to frequency and/or direction of the wave propagation determined by the angle θ. For instance, we can consider the incident angle θ as a control parameter for the frequency windows in $T(\omega)$, i.e., for the transparency edge in superconducting multilayers. For a layered sample with correlated disorder having the correlation function shown in Fig. 6b, the transparency versus the incident angle is shown in Fig. 6a. The filtering of THz radiation with respect to the incident angle is clearly seen.

4.2.6 Conclusion

We have shown that wave localization occurs for JPWs propagating in either disordered 1D Josephson arrays (1D Josephson transmission lines) or artificial superconducting multilayers. For samples with correlated disorder, transparency windows can be observed when varying the frequency or the angle of the incident electromagnetic waves. This transparency edge can be changed at will by an appropriate choice of the correlation function, which could be useful for THz devices, including frequency-selective or direction-selective filters. Also, the possibility to control the nonlinearity and dissipation in Josephson chains and arrays opens an avenue for experimental studies of these effects on localization. For example, it would be of interest to construct a transmission line with a quasi-periodic sequence of junctions where a singular continuous spectrum of states is realized. In this case, the power-law decrease of the transmission coefficient with an increase of the sample length should be observed [96]. Moreover, experimental and theoretical studies of JPW propagation in 2D Josephson arrays could be a useful tool to study wave localization in 2D structures.

4.3 SPECTRAL ANALYSIS AND SYNTHESIS OF CORRELATED SYSTEMS

Spectral characteristics of a large class of devices can be expressed in terms of the Fourier transform of the corresponding correlation function. Included in these systems are electronic nanodevices, optical fibers, rough surfaces, diffraction gratings, macromolecules, acoustic and electromagnetic waveguides. The propagation of surface plasmon polaritons [110], backscattering enhancement [24], and wave transport through narrow guiding surface-disordered systems [111] are sensitive to the roughness of the surface, which is mainly determined by the height- or gradient-roughness correlator. For a weak disorder, the pair correlator determines the transmission/reflection of classical or quantum waves through correlated random structures [25]. Diffraction gratings are one of the most bright and visible, in a literal sense, manifestations of the impact of correlations on the spectral properties of optical devices [112].

Disordered media have been studied since 1950s when the Anderson localization [113, 99], a fundamental concept for disordered systems, was proposed. A white-noise (uncorrelated) disordered 1D chain exhibits insulating behavior at low temperatures due to multiple random scattering, i.e., as a result of destructive interference, the quantum particle turns out to be localized inside a finite length (calculated by Thouless [100]). In contrast, particles in a periodic (i.e., perfectly ordered) system can propagate long distances, giving rise to metal-like conducting behavior or peculiar properties of photonic crystals [95]. Intermediate systems, described, for instance, by correlated random potentials (e.g., References [102, 103, 96]) or quasicrystalline structure (e.g., References [114, 115, 116]), can demonstrate either localization or delocalization, depending on their parameters. For instance, the degree of disorder drastically influences the particle mobilities. These phenomena involving disorder can be conveniently described using 1D tight-binding models (see, e.g., References [117, 118, 96, 114, 115]).

For a long time, it was believed that localized and extended states in 1D potentials did not coexist and 1D systems could not display complex dynamic features such as the metal-insulator transition. An unexpected result was obtained in References [106, 119, 120] that correlations can produce delocalized states in a random potential. Generalizing Thouless calculations, a relation between the pair correlation function and the localization length was derived in References [25, 102, 104, 105]. This relation shows that the correlation length can quickly change from infinite to a finite value at certain values of the particle energy; i.e., the localized and delocalized electron states do coexist in 1D random long-range-correlated systems. This means that the metal-insulator transition can be observed when the particle energy changes. This phenomenon, known as the mobility edge, is discussed, for instance, in Reference [23]. These findings have revised a previously accepted belief that any randomness (no matter how weak the randomness is) in 1D structures results in Anderson localization [99]. In contrast, long-range (i.e., power law decaying) correlations can provide a continuum of extended states in the energy spectrum, and a metal-insulator transition.

The problem of constructing a random sequence with predefined correlator has been partially solved in many independent papers (see, e.g., References [19, 20, 25, 121, 22, 27, 28]). However, the algorithm proposed in these articles generates correlated sequences of random variables whose values are not limited, i.e. may be taken on from $-\infty$ to ∞.

For practical applications it is desirable to generate a correlated disorder using only a *finite* number $\mathcal{N}$ of values of random elements. An example of such a system is the sequence of nucleotides in a DNA molecule, where Nature uses four elements to generate a random sequence with long-range correlations. It is not evident *a priori* whether a random sequence of only two elements may have an arbitrary correlation function, and what kind of restrictions the finiteness of the set of random values imposes. The challenging mathematical problem of constructing a *dichotomous (binari)*, $\mathcal{N}=2$, sequence with prescribed correlation properties was first considered in References [31, 107]. The theory developed in these papers was successfully applied to describe the correlation properties of coarse-grained literature texts Reference [107]. The importance of this issue may be explained also by the strong impact upon the creation of a new class of electronic, electromagnetic and optical devices, which can be used as window filters with unique transmission properties.

In this Subsection we apply the above-mentioned method [31, 107] for constructing a long-range correlated sequence of two-valued random elements $\epsilon(i)$ with a given correlator. The main idea of our work [122] is to construct the *binary* sequence of the required statistic properties and show effectiveness of the method. We illustrate this program for two simple, yet pronounced systems whose spectral properties are characterized by a pair correlation function. Let us stress the essential difference between two problems in the construction of the sequences when one of them is made from the discrete finite set of random values and another one is made from continuous infinite set.

In solving this problem we apply our instrument to generate the correlated binary sequence $\epsilon(i)$ with a prescribed correlation function $K(r)$ as an *additive* Markov chain. In order to introduce notations, we reproduce some definitions from Section 2. Consider a homogeneous binary sequence $\epsilon(i)\in\{\epsilon_0,\epsilon_1\}$, $i\in\mathbf{Z}=\{\ldots,-2,-1,0,1,2,\ldots\}$. To determine the N-step Markov chain we introduce the conditional probability function $P(\epsilon(i) \mid T_{N,i})$ of the additive form

$$P(\epsilon(i)=\epsilon_1 \mid T_{N,i}) = p_1 + \sum_{r=1}^{N} F(r)(\epsilon(i-r) - <\epsilon>). \tag{4.34}$$

As earlier, the memory function $F(r)$ describes the strength of impact of the previous symbols $\epsilon(i-r)$ $(r = 1, ..., N)$ upon a generated one, $\epsilon(i)$. The probability p_1 is the relative part of symbols ϵ_1 among all the symbols in the whole sequence.

The correlation properties of the sequence $\epsilon(i)$ is characterized by the pair correlation function and its normalized analog $\mathcal{K}(r)$,

$$K(r) = <\epsilon(i)\epsilon(i+r)> - <\epsilon(i)>^2, \qquad \mathcal{K}(r) = \frac{K(r)}{K(0)}. \tag{4.35}$$

The mean value of a function f of the variable $\epsilon(i)$ is

$$< f(\epsilon(i)) > = \lim_{M\to\infty} \frac{1}{2M+1} \sum_{i=-M}^{M} f(\epsilon(i)), \tag{4.36}$$

In other words, the ensemble average $<\cdot>$ is treated in the above equation as spatial for a stationary sequence $\epsilon(i)$, considered here.

A single-valued relation between the memory function $F(r)$ and the normalized correlation function $\mathcal{K}(r)$ is

$$\mathcal{K}(r) = (\epsilon_1 - \epsilon_0) \sum_{r'=1}^{N} F(r')\mathcal{K}(r-r'), \quad r \geq 1. \tag{4.37}$$

Equation (4.37) allows us to calculate the memory function $F(r)$ via correlator $\mathcal{K}(r)$ and construct the Markov chain effectively using the conditional probability $P(\cdot \mid \cdot)$ introduced in Eq. (4.34).

Now when the theoretical instrument of the Markov chains have been described, we can proceed with spectral analysis and synthesis of the binary sequences. The *analysis* consists of the calculation of the memory function via the correlator by solving Eq. (4.37). The generation of the binary sequence with the given memory function by Eq. (4.34) is the *synthesis*.

In various physical problems the most important characteristics can be expressed in terms of the Fourier transform of the correlation function. So, in order to determine a binary correlated sequence with a prescribed complexity, we design the Fourier transform $\hat{\mathcal{K}}(k)$ for all values of $k \in [-\pi, \pi]$ and calculate $\mathcal{K}(r)$ as inverse transform of $\hat{\mathcal{K}}(k)$,

$$\mathcal{K}(r) = \frac{1}{\pi} \int_0^{\pi} \hat{\mathcal{K}}(k) \cos(kr) dk. \tag{4.38}$$

After that we solve Eq. (4.37) for the memory function $F(r)$, and build the sequence of ϵ_0 and ϵ_1 using the conditional probability (4.34).

What kind of "building stone" can we use for this purpose? Suppose that $\hat{K}_p(k)$, $\hat{K}_{qp}(k)$, and $\hat{K}_{wn}(k)$ are the correlators of periodic, quasiperiodic crystals and amorphous solids

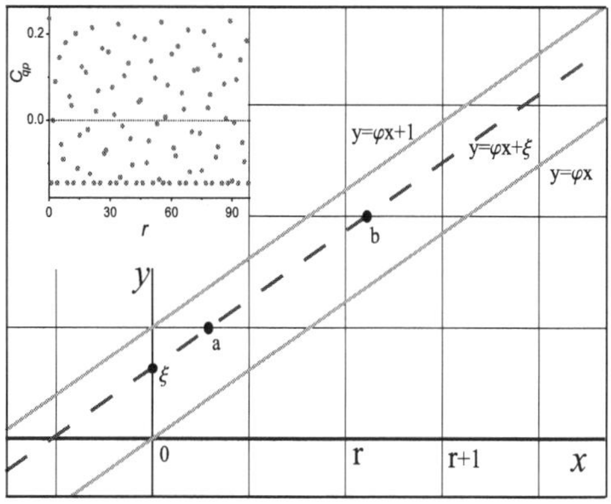

Figure 7 The line $y = \varphi x + \xi$ "generates" the Fibonacci sequence (4.42a). The strip bounded by two lines $y = \varphi x$ and $y = \varphi x + 1$ corresponds to an ensemble of possible Fibonacci sequences with $0 \leq \xi < 1$. The line $y = \varphi x + \xi$ intersects a horizontal line $y = 1$ between 0 and 1 (and so does another line y between r and $r + 1$) at point a (and b) and gives $\Phi_\xi(0) = 1$ (and $\Phi_\xi(r) = 1$). Such values of ξ give a nonzero contribution to the correlation function (4.43). The inset shows correlation function $K_{qp}(r)$ of the quasiperiodic Fibonacci sequence.

(white noise) respectively. We will try to construct the sequence with the following form of correlator,

$$\hat{K}(k) = \sum_i W_i(k)\hat{K}_i(k). \tag{4.39}$$

Here index i stands for one of three meanings p, qp or wn, and $W_i(k)$ are the weight functions.

According to the similarity of diffraction gratings and solids, each atom of a solid can be considered as a scatterer. In this way, after constructing a complex correlated grating, we could imagine that the design represents an artificial complex 1D solid. For the sake of illustration, we focus on two problems of general interest.

(A) In the whole interval of wave numbers $-\pi < k < \pi$, the correlator $\hat{K}(k)$ is taken in a form (4.39) with k-independent W_i. The constructed sequence $\epsilon(n)$ will then correspond to two types of atoms arranged in a global mixture of the ingredients.

(B) The interval of wave numbers $-\pi < k < \pi$ is divided into parts containing pure spectra of the above-mentioned solids divided by some relatively small transition region of mixed states. In this case functions $W_i(k)$ are smoothed step-wise ones. Such a construction is seen as a crystal, a quasicrystal or an amorphous solid versus the angle of observation α_s.

To complete the program and construct the total correlation function $K(r)$ of the design we have to calculate the "partial" ones and their Fourier transforms.

It follows from definition (4.35) that the correlation function $K(r)$ of a periodic sequence with identical values $\epsilon(n) = <\epsilon>$ is equal to zero for all distances r. Nevertheless, the normalized correlator is

$$\mathcal{K}_p(r) = 1, \qquad \hat{\mathcal{K}}_p(k) = \delta(k). \tag{4.40}$$

Here function $\delta(\cdot)$ is the Dirac delta-function.

The correlation function of an uncorrelated random sequence of $\epsilon(n)$ (or a white noise) is

$$\mathcal{K}_{wn}(r) = \delta_{r,0}, \qquad \hat{\mathcal{K}}_{wn}(k) = 1. \tag{4.41}$$

The Kronecker symbol is denoted as $\delta_{r,r'}$.

As a representative for a quasiperiodic sequence of $\epsilon(r)$, we consider the well-known Fibonacci chain defined as

$$\epsilon_\xi(r) = \epsilon_0 + (\epsilon_1 - \epsilon_0)\Phi_\xi(r), \qquad \xi \in [0, 1), \tag{4.42a}$$

$$\Phi_\xi(r) = [(r+1)\varphi + \xi] - [r\varphi + \xi], \tag{4.42b}$$

where the choosing of parameter ξ specifies a concrete sequence among all possible Fibonacci chains, $\varphi = (\sqrt{5} - 1)/2$ is the golden number and $[x]$ is the integer part of x. The geometrical meaning of this construction is explained in Fig. 7. In order to determine the correlation function of this sequence it is convenient to use an ergodic theorem of the homogeneity of the distribution of $y_n = n\varphi - [n\varphi]$ on the interval $[0, 1)$ for an arbitrary irrational number φ (see Ref. [123]) and calculate the correlation function $K(r)$ using the ensemble average as the integration of random function over the parameter ξ.

From the geometrical consideration in Fig. 7, we obtain the result

$$K_{qp}(r) = (\epsilon_1 - \epsilon_0)^2 \begin{cases} 2\varphi - 1 - \{\varphi r\}, & \{\varphi r\} \le 1 - \varphi, \\ 3\varphi - 2, & 1 - \varphi \le \{\varphi r\} < \varphi, \\ 2\varphi - 2 + \{\varphi r\}, & \varphi \le \{\varphi r\}, \end{cases} \tag{4.43}$$

where $\{x\}$ is the fractional part of x. The correlation function is defined for all integer arguments, which are incommensurable with period $1/\varphi$ of the function $\Phi_\xi(r)$; see the inset in Fig. 7. Numerical calculations of $K_{qp}(r)$ carried out using the spatial averaging exhibit good agreement with the analytical result (4.43).

Now that we know all the necessary correlators, we can design a compound correlation function as a linear combination of correlators of "pure" systems. So, let us proceed to solve the above-mentioned problems.

At first, as the realization of the problem (A), we consider *Fraunhofer diffraction* using the standard kinematic approach. We imagine the diffraction grating as a series of $2M + 1$ scatterers, equally spaced at a distance d with a scattering factor characteristic of the nth scatterer $\epsilon(n)$. The scattered waves resulting from interference form a wave diffraction pattern,

$$\mathcal{E}_M = \sum_{n=-M}^{M} \epsilon(n) \exp(i\varphi_n), \tag{4.44}$$

where the path length φ_n for the nth ray is

$$\varphi_n = kn, \qquad k = \frac{2\pi d}{\lambda}(\sin\alpha_i + \sin\alpha_s). \tag{4.45}$$

The direction of propagation of the incident wave is given by the angle α_i. The intensity of the scattered wave in the direction α_s is $|\mathcal{E}_M|^2$. By dividing this quantity by the number of scatterers, we obtain the normalized intensity per one scatterer,

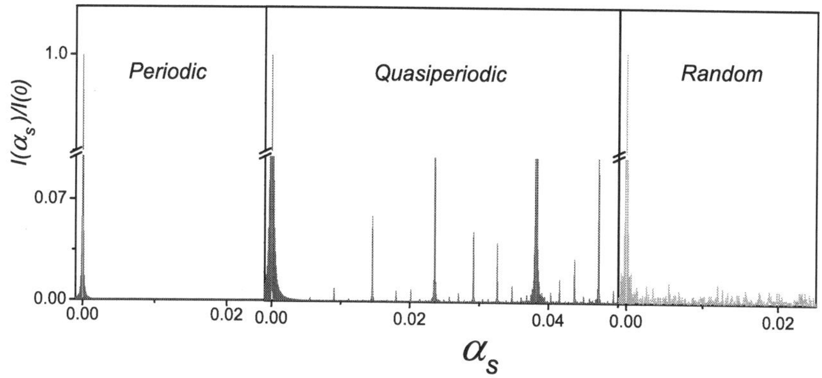

Figure 8 The intensity of scattered light vs. angle α_s obtained by numerical simulation for periodic, quasiperiodic and uncorrelated gratings with $\epsilon_0 = 0$ and $\epsilon_1 = 1$. The number of scatterers is 1000, $\lambda/d = 0.1$, and $\alpha_i = 0$. The small parts of the shown angles correspond to $k \in (0, \pi/2)$ for the quasiperiodic grating and $k \in (0, \pi/4)$ for the other two.

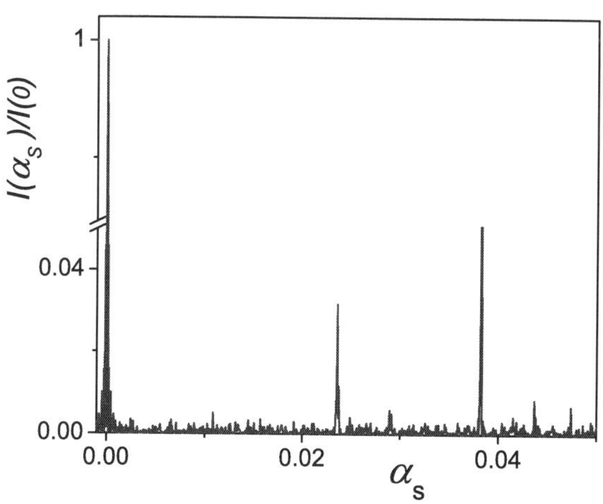

Figure 9 The intensity of scattered light vs. angle α_s obtained by numerical simulation for complex grating. The correlation function $\hat{K}(k)$ is made up of quasiperiodic (4.42) and uncorrelated (4.41) sequences, the correlators of which are mixed in proportion $W_p : W_{qp} : W_{wn} = 0 : 1 : 1$. The probabilities of scattering factors $\epsilon_0 = 0$ and $\epsilon_1 = 1$ occurring are equal to $1/2$. All other parameters coincide with those given in Fig. 8.

$$I(k) = \lim_{M\to\infty} \frac{|\mathcal{E}_M|^2}{(2M+1)} = I_p(k) + \hat{K}(k), \tag{4.46}$$

in the limit $M \to \infty$. The first contribution,

$$I_p(k) = 2\pi < \epsilon >^2 \sum_{n=-\infty}^{\infty} \delta(k - 2\pi n), \tag{4.47}$$

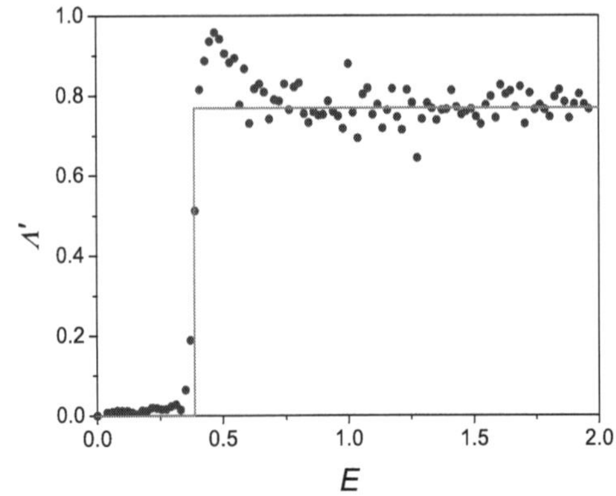

Figure 10 The normalized Lyapunov exponent $\Lambda'(E) = 32\ \sin^2(k)\ \Lambda_B(E)/(\varepsilon_1 - \varepsilon_0)^2$ vs. energy E. The solid line is the prescribed characteristic: $\Lambda' = 0$ for $0 < E < 0.4$ and $\Lambda' = 0.77$ for $0.4 < E < 2$. The dots are the result from the calculation of the Lyapunov exponent of the sequence $\varepsilon(n)$ constructed by means of function $F(r)$. The memory function is obtained by numerical solution of Eq. (4.37) for the memory length $N = 300$ and the correlation function $\mathcal{K}(r) = \sin(0.9\,\pi r)/(0.9\,\pi r)$, corresponding to the design.

to the intensity is that of an ideal periodic grating with all equal symbols $\epsilon(n) = <\epsilon>$, resulting from the Poisson summation formula.

The second term in Eq. (4.46) is the Fourier transform $\hat{K}(k)$ of the correlation function,

$$\hat{K}(k) = K(0) + 2\sum_{r=1}^{\infty} K(r)\cos(kr), \tag{4.48}$$

because of the quadratic form of the intensity in $\epsilon(n)$.

Figure 8 shows the intensity of waves scattered by the pure gratings, and Fig. 4.3 displays a numerical realization of the problem (A). This figure corresponds to a complex grating whose properties contain at the same time the features common to periodic, quasiperiodic and random gratings. We would like to stress the highly non-trivial character of the presented design.

The second problem (B) is illustrated by considering the stationary tight-binding Schrödinger equation

$$\psi_{n+1} + \psi_{n-1} + (E - \varepsilon(n))\psi_n = 0, \tag{4.49}$$

where E is the energy and $\varepsilon(n)$ is a binary site potential. Equation (4.49) is a prototype model for the description of propagation of excitations (e.g., electrons, photons and phonons) in one-dimensional lattices.

The correlations in the potential $\varepsilon(n)$ determine the energy spectrum of the system. The general decomposition of the spectral measure (or the integrated density of states) [124] is analagous with that of the intensity measure. For an uncorrelated random potential all eigenstates are exponentially localized (Anderson localization [99]), which manifests in a Lyapunov exponent $\Lambda > 0$, and the spectrum has a pure point character. The translational symmetry of a periodic potential implies extended ($\Lambda = 0$) Bloch states and an absolutely continuous spectrum. The Fibonacci potential shows an exotic situation in that the spectrum is singular continuous [125], and all wave functions are (non-normalizable) "critical". The question of co-existence of the different energy spectra is an intriguing one.

In the Born approximation ($|\varepsilon_1 - \varepsilon_0| \ll 1$) the Lyapunov exponent (often equal to the inverse localization length) can be expressed [102, 104, 25] in terms of the Fourier transform $\hat{\mathcal{K}}$ of the normalized correlation function as

$$\Lambda_B(E) = \frac{(\varepsilon_1 - \varepsilon_0)^2 \hat{\mathcal{K}}(2k)}{32 \sin^2 k}, \quad k \in [-\pi, \pi]. \tag{4.50}$$

In the zero approximation on the strength of disorder the energy relates to the wave number k by the formula $E = -2\cos k$. Equation (4.50) reveals a relation between correlations $\mathcal{K}(r)$ in the site potential and localization properties of the eigenstates. This observation enables the construction of binary correlated sequences $\varepsilon(n)$ with prescribed spectral characteristics and provides a recipe for designing filters of an different complexity.

The example shown in Fig. 10 provides a demonstration of the problem (B) by designing a stepwise normalized Lyapunov exponent $\Lambda'(E) = \hat{\mathcal{K}}(2k)$ in a region of energy E (or wave number k), which defines a sharp mobility edge within the second order perturbation theory.

In conclusion, we have applied the method to construct binary sequences with predefined correlation properties, and designed a complex grating and an approximative mobility edge in one dimension as theoretical realizations of the program. The effectiveness of the method has been demonstrated by the numerical simulation. Besides, the analytic expression for quasiperiodic correlator has been obtained.

4.4 RANK DISTRIBUTIONS AND THE ZIPF LAW

The rank distributions in stochastic systems attract the attention of specialists in physics and many other fields of science because of their universal power-law character, $W \propto R^{-\alpha}$ (the so-called Zipf law [6]). Here W is the frequency of some object occurring and R stands for the rank of this object. Discovered originally for the rank distribution of words in the natural languages, the Zipf law was later observed in the rank distributions of other objects, such as distributions of "words" in the DNA sequences [7], PC codes [7], seismology [13], capitals of stock market players [126] (in economics, the Zipf law in slightly different form is known as Pareto's principle or the 80–20 rule [11]), the population of cities, areas occupied by countries, masses of cosmic objects etc (see [14], [127]). In spite of numerous attempts to describe this phenomenon analytically, a deep insight into the problem has not so far been gained. In the strict sense, the Zipf law implies the closeness of the exponent α to unity. Besides, the Zipf law should be valid within the fitting range of about 2–3 decades. From this point of view, the rank distributions in some above-mentioned real systems do not comply with these criteria. Therefore, we will use the term "Zipf's-like law" for the power rank distributions if α differs noticeably from unity.

To define the rank distribution $W(R)$ of some objects in a definite sequence, it is necessary to establish a correspondence between objects and their frequencies of appearing in a sequence and to arrange the objects in ascending order of these frequencies. Analyzing the rank distribution in a dynamic system we could find ourselves faced with two possibilities. The first one is characterized by the power-law rank distribution as a result of some kind of probabilistic game established by *a priori* defined non-equivalence of words. The

Mandelbrot models [128, 129], Li model [130], and many other models are constructed on the basis of the choice of the *a priori* nonequivalent competitors, and specifically this non-equivalency is a reason for obeying the rank distributions to the Zipf law. For example, the non-equivalency of the words in literary texts is caused by their different lengths and, consequently, by their different statistical weight that is defined by the number of characters in the word and the capacity of an alphabet. To illustrate this we will consider an elementary example below. In the second possibility, the power-law rank distribution appears as a result of correlations in a system. In this case, the competitive objects are considered as *a priori* equivalent, i.e. having "the same rights" in the competition [5]. The rank distribution of the overlapping triplets in the DNA sequences can serve as a vivid example of the real systems for which this approach is essential (see [7]).

We suggest an analytically solvable model of the additive many-step Markov chain where the rank distribution of different L-words (L consequent symbols in the chain) is examined. In order to increase the statistical sampling of the words, we consider all L-words in the chain, that could overlap each other. A special numerical analysis shows that such a type of sampling does not effect the rank distribution of the L-words. Since the words are equal in their length, the rank distribution in this system occurs as a result of the correlations. In other words, our model is based specifically on the second approach to the choice of the competitors in the sequence.

We have analytically studied the rank distributions of words of the length L in Markov chains. If a Markov chain possesses the step-like memory function considered in Ref. [32], this distribution is shown to be of the staircase form. For strong correlations, the envelope curve of the rank distribution obeys the power-law with the exponent of the order of unity, i.e. the distribution is described by the Zipf's-like law. The obtained results allow us to express our view upon the origin of Zipf's law. In particular, we have made sure that the correlations of symbols within the competitive words is sufficient for the appearance of the Zipf's-like law in their rank distributions. Contrary to the speculations about the connection of the Zipf's-like law to the *long-range correlations* (with the scale of about the text length) that were expressed clearly by a number of authors [131, 21], we have demonstrated that the *short-range correlations* (with the scale of about the word length L) can result in such a rank distribution.

The suggested approach to the problem of the Zipf law is expedient because we are provided with theoretical parameters that affect both the character of correlations and the rank distribution of words occurring in the Markov chain. Due to this circumstance, we examine the relationship between the rank distributions and the correlation properties of the system.

4.4.1 Elementary example

This example is actually similar to the model developed by W. Li in Reference [130]. Let us consider a homogeneous stationary unbiased binary sequence of symbols, $a_i = \{0, 1\}$. The L-word is defined as a set of sequential symbols of definite length L being taken at the arbitrary point of a sequence. The rank distribution of words is a relationship connecting the probability W of a certain word occurring to the corresponding rank. The words are ordered in ascending rank order, $W(1) \geq W(2) \cdots \geq W(2^L)$. This model represents an example of a system where the competitors have *a priori* non-equivalent "rights".

Consider a set of words of arbitrary lengths in a non-correlated binary sequence including all one-symbol words, all two-symbol words, all three-symbol words, and so on up to some finite length $L_0 \gg 1$. The frequency of a concrete L-word appearing in the non-correlated sequence is evidently proportional to 2^{-L},

$$W(L) \propto 1/2^L \tag{4.51}$$

and does not depend on the arrangements of symbols in the word. This means that the frequency-rank plot is degenerated, i.e. it consists of steps, each of them corresponding to the set of words with the fixed length L. Let us derive a function $W(R)$ corresponding to the envelope curve passing through the right edges of each step on the plot. It is obvious that the length of the step corresponding to the L-words equals 2^L, so the rank of the right edge point of this step is described by the equation,

$$R(L) = \sum_{i=0}^{L} 2^i. \tag{4.52}$$

We can easily derive the function W(R) from Eqs. (4.51), (4.52),

$$W(R) = \frac{\text{Const}}{R+1}, \tag{4.53}$$

i.e. the Zipf distribution.

In the next Section, to avoid this trivial situation, we consider the ranking of the words of equal length L, i.e. to perform the so-called n-tuple Zipf-analysis [7] of a binary system. It is obvious that in the case of uncorrelated sequences the distribution is flat, i.e. all words have equal rank frequencies. Therefore, the correlations must be introduced into the systems under consideration.

4.4.2 Model

We examine the binary *N-step Markov chain*, which is determined by the *conditional probability* $P(a_i \mid a_{i-N}, a_{i-N+1}, \dots, a_{i-1})$ of the definite symbol a_i occurring (for example, $a_i = 0$) after symbols $a_{i-N}, a_{i-N+1}, \dots, a_{i-1}$

$$P(a_i = 0 \mid a_{i-N}, a_{i-N+1}, \dots, a_{i-1}) = P_k = 1/2 + \mu(1 - 2k/N). \tag{4.54}$$

Here μ is the strength of correlations in the sequence, $-1/2 < \mu < 1/2$. We will consider the case of $\mu \geq 0$ only, the formalism in the opposite case $\mu < 0$ has some distinctions.

Let us introduce a probability $b_N(a_1 a_2 \dots a_N)$ of the certain N-word $(a_1, a_2, \dots, a_N)$ occurring. We could easily verify that b_N satisfies the condition of compatibility for the Chapmen-Kolmogorov equation (see, for example, Reference [60]):

$$b_N(a_1 \dots a_N) = \sum_{a=0,1} b_N(a a_1 \dots a_{N-1}) P(a_N \mid a, a_1, \dots, a_{N-1}). \tag{4.55}$$

Here we have 2^N homogeneous algebraic equations for 2^N probabilities b_N of the N-words occurring. Besides, the normalization equation $\sum b_N = 1$ should be satisfied.

In the case under study, the set of equations can be substantially simplified owing to the following statement. It was proved [58] that the probability $b_N(a_1 a_2 \ldots a_N)$ depends on the number k of unities in the N-word only, i. e., it is independent of the arrangement of symbols in the word $(a_1, a_2, \ldots, a_N)$. This statement is valid because of the chosen simple model (4.54) of the Markov chain.

In Reference [58], Eqs. (4.55) were solved and then the distribution function of words with arbitrary length $L \leq N$ was derived,

$$b_L(k) = b_L(0) \frac{\Gamma(n+k)\Gamma(n+L-k)}{\Gamma(n)\Gamma(n+L)}, \qquad 0 \leq k \leq L \leq N, \tag{4.56}$$

with

$$b_L(0) = \frac{4^n}{2\sqrt{\pi}} \frac{\Gamma(1/2+n)\Gamma(n+L)}{\Gamma(2n+L)}.$$

This formula defines the probability of the L-word containing k unities occurring with a definite arrangement of symbols. This probability depends on the number k but does not depend on the arrangement of symbols in the word. The parameter $n > 0$,

$$n = \frac{N(1-2\mu)}{4\mu}, \tag{4.57}$$

governs the strength of the persistent correlations. The case $n \to \infty$ corresponds to the absence of correlations. Under this situation, the probability b_L does not depend on k. At $n \to \infty$, Eq. (4.56) yields $b_L = 1/2^L$ that coincides with the frequency (4.51) of the L-words occurring in the non-correlated sequence. The case $n \ll 1$ conforms to the strong persistence. In the next Subsection, we make use of Eq. (4.56) to derive the rank distribution of L-words in the Markov chain under consideration.

4.4.3 Zipf's-like distribution for $L \leq N$

At finite values of $n > 0$, the probability b_L depends on k. A simple analysis shows, that the plot $b_L(k)$ has a concave form because of the persistence ("attraction" of symbols to each other) in the system. The words in which unities (or zeros) are prevalent occur more frequently than the words with an approximately equal number of different symbols. Specifically this property provides the non-flat rank distribution of words with equal lengths in the Markov chain.

In order to clarify the form of the rank distribution, we note that the specific degeneration of the frequency of word occurrence takes place. The probability $b(k)$ (here and below we omit the subscript L) does not depend on the arrangement of symbols within the L-word. This means that the rank distribution of the L-words with $L \leq N$ should be of a staircase form. Yet another kind of degeneration results from the non-bias property of the sequence: the probability $b(k)$ is symmetric with respect to the change $k \leftrightarrow (L-k)$, $b(k) = b(L-k)$. Thus, as a result of these two symmetries, $2\mathrm{C}_L^k = 2L!/k!(L-k)!$ different words occur with the same frequency $b(k)$. Each of the steps can be labelled by the number $k \leq L/2$ of unities (or zeros) within them and is characterized by the length equal to the degeneracy multiplicity $2\mathrm{C}_L^k$. The right edge of the kth step corresponds to the rank $R_{max}(k)$,

$$R_{max}(k) = 2\sum_{i=0}^{k} C_L^i. \tag{4.58}$$

The ranks of all words containing k unities belong to the interval

$$R_{max}(k-1) + 1 \leq R \leq R_{max}(k). \tag{4.59}$$

Equations (4.56) and (4.58) being considered as a parametrically defined function $W(R) = b(k(R))$ yields the envelope curve passing through the right edges of the steps in the rank-distribution.

Using the Stirling formula for the Gamma-functions (which is valid at L, k, $(L-k) \gg 1$) and replacing the summation in Eq. (4.58) by integration, we can easily obtain the asymptotic expression for the dependence $R(W)$,

$$R = 2^L \sqrt{\frac{\zeta}{\pi}} \left(\frac{W}{B}\right)^{-\frac{1}{\zeta}} \ln^{-1/2}\left(\frac{W}{B}\right), \qquad \zeta = \frac{1}{1+2n/L} \tag{4.60}$$

with

$$B = 4^n \sqrt{\pi} \frac{\Gamma(1/2+n)}{\Gamma(n)\Gamma(2n+L)} \left(n + \frac{L}{2} - 1\right)^{2n+L-1} \exp(-2n - L + 2). \tag{4.61}$$

The distribution Eq. (4.60) differs from the power-law form by the logarithmic multiplier. If we neglect this weak logarithmic dependence, the Zipf's-like law for the rank distribution would be obtained from Eq. (4.60),

$$W \propto R^{-\zeta}. \tag{4.62}$$

Actually, the change (4.58) of variable, $k \to R(k)$, i.e. stretching the abscissa axis, slows down the decay of the frequency W with an increase of k and results in the power-law Eq. (4.62). A similar result was achieved in Subsection 4.4.2 by the change of variable $L \to R(L)$ Eq. (4.52) for the rank distribution of the words with different lengths L in the non-correlated sequence. Such a procedure was first outlined in Reference [130].

The obtained result is demonstrated in Fig. 11. The dotted line shows the plot of the rank distribution obtained from Eqs. (4.56) and (4.58), (4.59) at $L = 14$, $N = 15$, $\mu = 15/46$, $n = 4$. This plot passes closely to the solid line, which demonstrates the results from the numerical simulations of the rank distribution of words having the length $L = 14$ in the Markov chain generated using the same parameters N and μ [1]. The dash-dotted line in this figure is the envelope curve described by Eq. (4.60). The exponent in Zipf's distribution (4.62) is governed by the ratio n/L. In the case of weak correlations, at $n \to \infty$, the value of ζ tends to zero and the Zipf distribution appears to be destroyed, i.e. all the words of the length L occur with almost the same probabilities $W = 2^{-L}$. The opposite situation, $n/L \to 0$, corresponds to the strong correlations in the Markov chain. Equation (4.61) shows

[1] The first stage in the construction of the N-step Markov chain is a generation of the initial non-correlated N symbols, zeros and unities, identically distributed with equal probabilities 1/2. Each consequent symbol was then added to the chain with the conditional probability determined by the previous N symbols in accordance with Eq. (4.54)

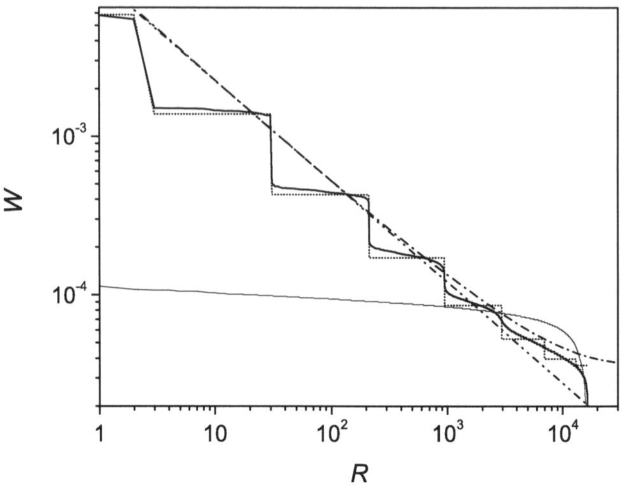

Figure 11 The rank distribution $W(R)$ of the words with length $L=14$ in the Markov chain determined by the conditional probability Eq. (4.54) at $N=15$, $\mu=15/46$. The solid line is for the numerical simulations, the dotted line describes the exact distribution obtained from Eqs. (4.56), (4.58) and (4.59), the dash-dotted line is the plot of envelope asymptotics (4.60), the dash-dot-dotted line describes the Zipf's-like asymptotics Eq. (4.62) with $\zeta=7/11$. The thin solid line obtained numerically at $\mu=-15/46$, $N=15$, $L=14$ is for the anti-persistent correlations.

that in this case the exponent ζ tends to unity. The plot of Zipf's distribution (4.62) with $\zeta=7/11$ is demonstrated by the dash-dot-dotted line in Fig. 11.

4.4.4 Scaling property of the rank distribution

The Zipf's-like law is frequently associated with the property of scale invariance [21], i.e. the invariance of the exponent in the Zipf law with respect to a certain decimation procedure. An analogous property referred to as the self-similarity appears in the frame of the model presented above. Let us reduce the N-step Markov sequence by regularly (or randomly) removing some symbols and introduce the decimation parameter $\lambda < 1$ which represents the fraction of symbols kept in the chain. As shown in Reference [58], the reduced chain possesses the same statistical properties as the initial one but is characterized by the renormalized memory length, N^*, and the persistence parameter, μ^*,

$$N^* = N\lambda, \qquad \mu^* = \mu\frac{\lambda}{1-2\mu(1-\lambda)}. \tag{4.63}$$

Indeed, the conditional probability P_k^* of the symbol zero after k unities occurring among the preceding N^* symbols in the reduced chain is described by Eq. (4.54) where N and μ should be replaced by renormalized parameters N^* and μ^*. Considering the Zipf law, we are interested in the invariance of the Zipf plot with respect to the decimation procedure. According to Eqs. (4.60) and (4.61), the slope of the Zipf plot depends on the parameter n only. This parameter remains unchanged after the transformation $N \to N^*$, $\mu \to \mu^*$ (see Eqs. (4.57) and (4.63)). As a result, the Zipf plots for rank distributions of the L-words with

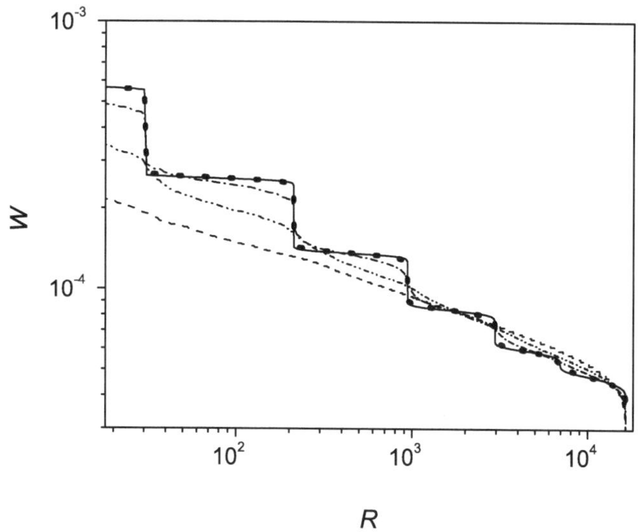

Figure 12 Rank distributions of the L-words with $L = 14$ in the N-step Markov sequences reduced by randomly removing some symbols. The solid line is for the initial sequence with $N = 32$, $n = 15$. The symbols corresponding to the decimation parameter $\lambda = 2$ ($N^* = 16$) lie almost on the solid one. Other lines are for decimated sequences with $N^* < L$. Specifically, the dash-dotted, dash-dot-dotted, and dashed lines correspond to $N^* = 8$, $N^* = 4$, and to $N^* = 2$, respectively.

$L < N$, obtained from the initial and decimated sequences, coincide. This self-similarity property is illustrated in Fig. 12.

The self-similarity is often observed in the correlated symbolic systems. However, this property is not necessary for the appearance of Zipf's-like rank distribution. Subsection 4.4.1 and papers [130] provide examples of such systems.

4.4.5 Rank distribution for $L > N$. Discussion

Now, let us look into the rank distribution of L-words with $L > N$. This problem is not amenable to analytic calculations, and, therefore, numerical simulations are applied. In this case, the above-mentioned degeneration of the probability of word occurrence does not exist. So, smoothing of the steps in the rank distribution takes place at $L > N$. This smoothing occurs gradually as the word length L increases, and the steps appear to be completely smoothed away at sufficiently high values of L (see curves in Fig. 13). This means that contrary to the case $L \leq N$, the rank distribution itself has the Zipf's-like form.

We would like to draw attention to the non-monotonous behavior of the Zipf slope ζ with an increase in the word length L. As seen from Eq. (4.61), this parameter increases at $L < N$. This growth continues at $L > N$ as well but only up to a certain value of $L = L_{cr} > N$. The maximum of ζ appears at $L = L_{cr} > N$ and then, at $L > L_{cr}$, ζ starts to decrease. This decrease is demonstrated in the inset to Fig. 13.

It is necessary to note that the position $L = L_{cr}$ of the maximum in the $\zeta(L)$ dependence is strongly related to the characteristic correlation length $2l_c + N$ in the Markov chain being studied. According to Reference [58], the symbols correlate with each other not only within the memory length N but within the enlarged region $2l_c + N$ where l_c represents the

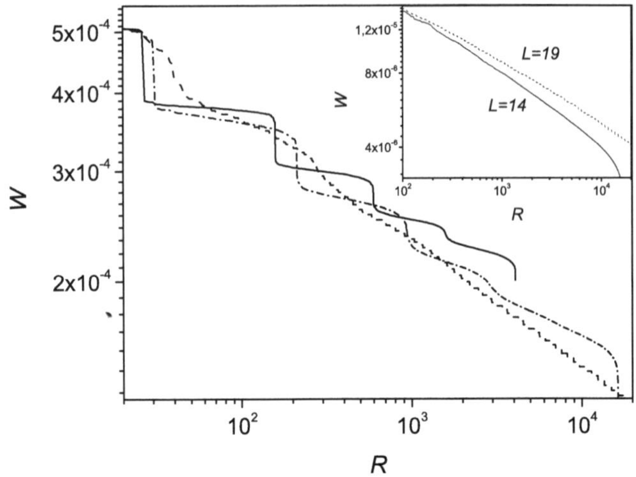

Figure 13 The Zipf plots for L-words with $L > N$. The solid, dash-dotted, and dashed lines are for $L = 12$, 14, 18, respectively. The parameters of the Markov chain are $N = 12$, $\mu = 0.1$. The phenomenon of step smoothing is noted with a growth in the word length L. In the inset: the Zipf plots for L-words with $N = 4$, $\mu = 0.1$ at $L > L_{cr}$ (the lengths of words are shown near the curves).

characteristic attenuation length of the fluctuations. Thus, the highest value of the exponent ζ in the rank distribution of words in the Markov chain is achieved if the size of the competitive words is close to the correlation length.

If the words are shorter than the correlation length, $L < N + 2l_c$, the specific features of the correlations become apparent in the rank distribution that results in deviations from the power-law. In the system that is considered in this volume, the deviations from the power-law manifest themselves in the occurrence of the steps in the rank distribution and in the additional weak logarithmic multiplier in Eq. (4.60). Moreover, at $N \sim 1$ the rank distribution deviates significantly from the power-law and assumes the exponential shape at $N = 1$ [21]. In the opposite limiting case, at $L \gg N + 2l_c$, the correlations over the whole word length disappear, and the rank distribution tends to a constant.

Note that specifically the persistent correlations in the Markov chain (that correspond to the attraction between the same symbols) lead to the pronounced Zipf law in the rank distribution of words. Indeed, the thin solid line in Fig. 11 demonstrates the very weak $W(R)$ dependence for the case of the antipersistent correlations, at $\mu = -15/46$.

4.4.6 Conclusion

We have presented a model of a symbolic stochastic dynamical system with non-zero correlations possessing the power-law rank distribution with the exponent of the order of unity. Direct analytic connection between the correlation parameters of the Markov chain and the power-law rank distribution is established. The remarkable property of invariance with respect to the decimation procedure for words shorter than the memory length is observed. It is shown that the Zipf's-like law occurs if the length of the competitive words is of the order of the correlation length.

The obtained results allow us to suggest the following physical picture of the manifestation of the Zipf's-like law in the correlated systems. The correlations should be presented in the system but the noise of sufficient intensity should be imposed on these correlations over the length of competitive words. Within the considered stochastic system, this noise is provided by sufficiently strong fluctuations observed on the scales of the word length. The role of the noise consists of concealing the specific peculiarities of the correlations. Owing to the fluctuations, the concrete shape of the correlations does not appear to be very important. Accordingly, the Zipf law is a consequence of the rather rapidly damping persistent correlations of quite arbitrary form, i.e. the global correlations in the system are not necessary. The Zipf law is a manifestation of the inner microstructure of the system and as a result of attraction between building blocks of the same kind.

We would like to emphasize that both circumstances, the difference in the length of words and the correlations between symbols, could be of great significance in forming the Zipf law in the rank distribution of words in correlated symbolic systems. Special investigations are necessary to clarify the role of these factors in the rank distributions in concrete real systems.

4.5 TWO KINDS OF CORRELATIONS IN LITERARY AND DNA "TEXTS"

In this Subsection, the binary multi-step Markov chain with the additive memory function is considered as a model for the analysis of the coarse-grained naturally written texts. They are shown, in References [58, 107], to be strongly correlated sequences that possess anti-persistent correlations at small distances (in the region of grammatical rules action, $L < 300$) and persistent ones at long distances (in the region of semantic rules action, $L > 300$). Our analysis shows that the whole literary work is self-similar globally correlated system. The last point of the Bible "remember" its first letter!

For some concrete examples of literary texts, the memory functions are obtained and their power-law behavior at long distances is disclosed. This property is shown to be a cause of self-similarity of texts with respect to the decimation procedure.

4.5.1 Introduction

Important examples of symbolic stochastic dynamic systems with long-range correlations are literary texts. The statistical properties of the coarse-grained texts written in any language exhibit a remarkable deviation from random sequences [132, 5, 133, 134]. In this Subsection, we examine the coarse-grained literary texts as the additive Markov chains. First, we apply the theory of Markov chain with step-wise memory function developed in Subsection 2.1. Next, we use the obtained above equation (2.109) that connects mutually-complementary characteristics of these sequences, the memory and correlation functions. Once the memory function of the original sequence is found from the analysis of the correlation function, we construct the corresponding Markov chain with the same statistical properties.

We show that the memory function of any coarse-grained literary text is characterized by a complex structure because of the competition between two kinds of correlations. One type of correlations is efficient at short distances, $L \lesssim 300$. The corresponding MF is negative, which reflects the *anti-persistent* nature of such correlations. The other type of correlations with positive memory function act at long distances, $L \gtrsim 300$. The strength of these *persistent* correlations decreases as a power-law function. We demonstrate that the power-law decrease of the memory function results in the self-similarity phenomenon in the coarse-grained texts with respect to the decimation procedure.

4.5.2 Step-wise memory function's approach

In order to check the applicability of the theory of the binary N-step Markov chains to literary texts we resorted to the procedure of coarse graining by the random mapping of all characters of the text onto the binary set of symbols, zeros and unities. The statistical properties of the coarse-grained texts depend, but not significantly, on the kind of mapping. This is illustrated by the curves in Fig. 14 where the variance $D(L)$ for five different kinds of the mapping of Bible is presented. In general, the random mapping leads to nonequal numbers of unities and zeros in the coarse-grained sequence. Analysis carried out in Reference [59] indicates that the variance $D(L)$, see Eq. (2.98) gets the additional multiplier, $1-4\nu^2/(1-2\mu)^2$, in the biased case. In order to derive the function $D(L)$ independent of the relative number of unities and zeros, we divided the numerically calculated value of the variance by this multiplier.

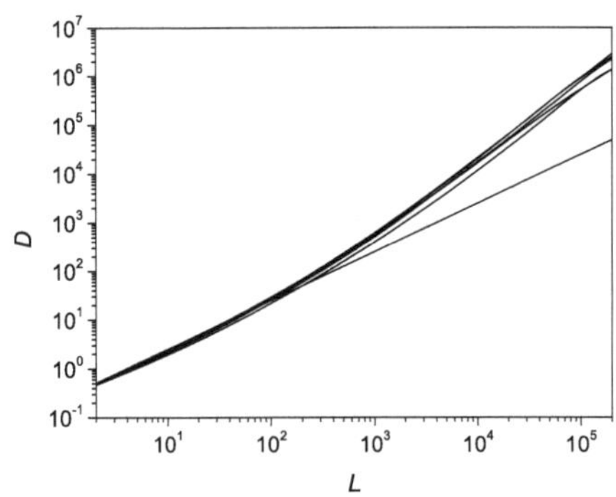

Figure 14 The dependence $D(L)$ for the coarse-grained text of Bible obtained by means of five different kinds of random mapping. The straight solid line describes the non-correlated Brownian diffusion, $D(L)=L/4$.

The study of different written texts has suggested that all of them are characterized by the pronounced persistent correlations. It is demonstrated by Fig. 15 where the five variance curves go significantly higher than the straight line $D=L/4$. However, it should be emphasized that regardless of the kind of mapping the initial portions, $L<80$, of the curves correspond to a slight anti-persistent behavior (see inset to Fig. 16). Moreover, for some inappropriate kinds of mapping (e.g., when all vowels are mapped onto the same symbol) the anti-persistent portions can reach the values of $L \sim 1000$. To avoid this problem, all the curves in Fig. 15 are obtained for the definite representative mapping: (a-m) $\rightarrow$ 0; (n-z) $\rightarrow$ 1.

Thus, the persistence is the common property of the binary N-step Markov chains that have been considered in this volume and the coarse-grained written texts at large scales.

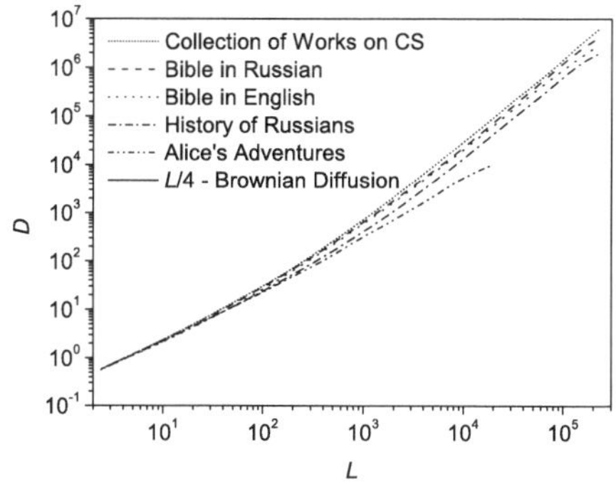

Figure 15 The dependence $D(L)$ for the coarse-grained texts of collection of works on computer science ($m = 2.2 \cdot 10^{-3}$, short-dotted line), Bible in Russian ($m = 1.9 \cdot 10^{-3}$, dashed line), Bible in English ($m = 1.5 \cdot 10^{-3}$, dotted line), "History of Russians in the 20-th Century" by Oleg Platonov ($m = 6.4 \cdot 10^{-4}$, dash-dotted line), and "Alice's Adventures in Wonderland" by Lewis Carroll ($m = 2.7 \cdot 10^{-4}$, dash-dot-dotted line). The solid line describes the non-correlated Brownian diffusion, $D(L) = L/4$.

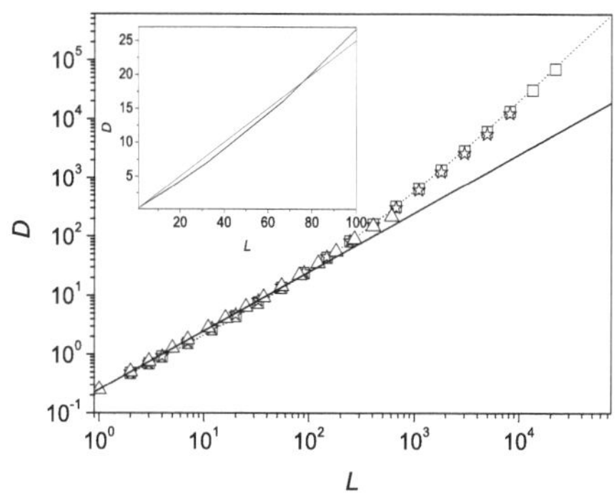

Figure 16 The dependence of the variance D on the tuple length L for the coarse-grained text of Bible (dotted line) and for the decimated sequences with different parameters λ: $\lambda = 3/4$ (squares), $\lambda = 1/2$ (stars), and $\lambda = 1/256$ (triangles). The solid line describes the non-correlated Brownian diffusion, $D(L) = L/4$. The inset demonstrates the anti-persistent portion of the $D(L)$ plot for Bible.

Moreover, the written texts as well as the Markov sequences possess the property of self-similarity. Indeed, the curves in Fig. 16 obtained from the text of Bible with different levels of deterministic decimation demonstrate self-similarity. Presumably, this property is the mathematical reflection of the well-known hierarchy in the linguistics: *letters* $\to$ *syllables* $\to$ *words* $\to$ *sentences* $\to$ *paragraphs* $\to$ *chapters* $\to$ *books*.

All the above-mentioned circumstances allow us to suppose that our theory of the binary N-step Markov chains can be applied to the description of the statistical properties of the texts of natural languages. However, in contrast to the generated Markov sequence (see Fig. 17) where the full length $\mathcal{M}$ of the chain is far greater than the memory length N, the coarse-grained texts described by Fig. 15 are of relatively short length $\mathcal{M} \lesssim N$. In other words, the coarse-grained texts are similar not to the Markov chains but rather to some non-stationary short fragments. This implies that each of the written texts is correlated throughout the whole of its length. Therefore, as far as the written texts are concerned, it is impossible to observe the second portion of the curve $D(L)$ parallel (in the log-log scale) to the line $D(L) = L/4$, similar to that shown in Fig. 17. As a result, we cannot define the values of both parameters N and μ for the coarse-grained texts. The analysis of the curves in Fig. 15 can give the combination $m = 2\mu/N(1-2\mu)$ only (cf. Eq. (4.57)). Perhaps, this particular combination is the real parameter governing the persistent properties of the literary texts.

We would like to note that the origin of the long-range correlations in the literary texts is hardly related to the grammatical rules as is claimed in Reference [5]. At short scales $L \leq 80$ where the grammatical rules are in fact applicable the character of correlations is anti-persistent (see the inset in Fig. 16) whereas semantic correlations lead to the global persistent behavior of the variance $D(L)$ throughout the entirity of literary text.

The numerical estimations of the persistent parameter m and the characterization of the languages and different authors using this parameter can be regarded as a new intriguing problem of linguistics. For instance, the unprecedented low value of m for the very inventive work by Lewis Carroll as well as the closeness of m for the texts of English and Russian versions of Bible are of certain interest.

4.5.3 Two kinds of correlations in literary texts

In this Subsection we apply a more sophisticated tool to investigating the correlation properties of the coarse-grained literary texts than in previous Subsection. We will use here the results of Subsection 2.2 without the step-wise memory function approximation.

Variance and correlation function. At the outset of this Subsection, we examine the variance $D(L)$ of the coarse-grained text of the King James Version of the Bible [135]. The result from the numerical simulation is represented by the solid line in Fig. 17. The straight dotted line describes the variance $D_0(L) = L\bar{a}(1-\bar{a})$, which corresponds to the non-correlated biased Brownian diffusion. One of the typical coarse-graining procedure was used to map the letters of the text onto the symbols zero and unity, $((a-m) \mapsto 0$, $(n-z) \mapsto 1)$. It is clearly seen that the diffusion is antipersistent at small distances, $L \lesssim 300$, (see inset) whereas it is persistent at long distances[2]. The deviation of the solid line from the dotted one testifies to the existence of the correlations in the text of the Bible. To confirm this statement we break down the original text into subsequences of a given length $L_0 = 3000$ and randomly shuffle them. The results from the calculation of the variance for the coarse-grained initial and shuffled texts of the Bible are given in Fig. 18. For $L \ll L_0$, the difference in $D(L)$ is negligible. At $L \sim L_0$, the variance and correlation function of the shuffled sequence are less than the original ones. At $L \gg L_0$, the correlations in the shuffled text vanish. In this region, the variance $D(L)$ is a linear function, and the correlation function being the second derivative of the variance equals to zero.

It is easy to show that the correlation function of the shuffled sequence can be written as,

$$K(r) = \begin{cases} K_0(r)(1-\frac{r}{L_0}), & r < L_0, \\ 0, & r \geq L_0, \end{cases} \tag{4.64}$$

[2] At length L of the order of the full text length $\mathcal{M} \sim 10^6$, we observe a violation of the monotonous growth of the variance $D(L)$. It is a manifestation of the border effect. Obviously, $k(L) \to \bar{k}$ and the variance $D(L) \to 0$ (see definition Eq. (2.9)) at $L \to \mathcal{M}$.

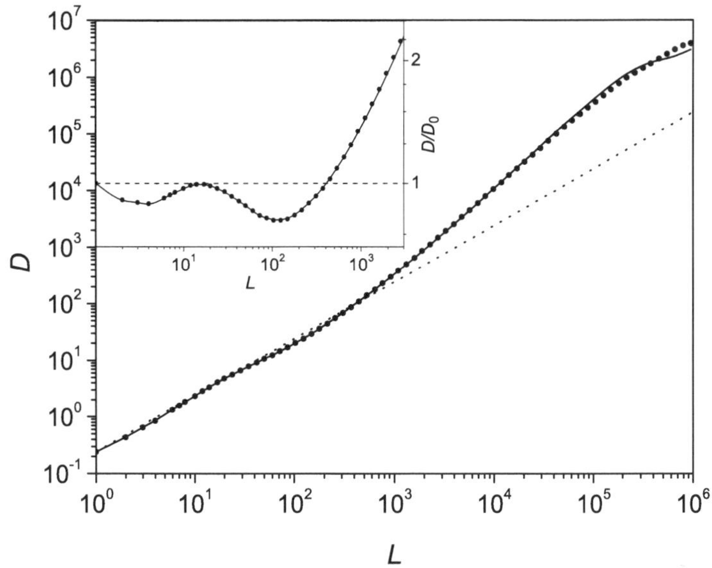

Figure 17 The variance $D(L)$ for the coarse-grained text (letters $(a-m) \mapsto 0$, letters $(n-z) \mapsto 1$) of the Bible (solid line) and the Markov chain generated by means of the reconstructed memory function $F(r)$ (filled circles). The dotted straight line describes the non-correlated Brownian diffusion, $D_0(L) = L\bar{a}(1-\bar{a})$. The inset demonstrates the antipersistent dependence of the dimensionless ratio $D(L)/D_0(L)$ upon L at short distances.

where $K_0(r)$ is the correlation function of the original non-shuffled sequence. The corresponding variance obtained by double numeric integration (see Eq. (1.11)) of the function $K(r)$ given by Eq. (4.64) is shown in Fig. 18 by the solid line.

Along with the global characteristic $D(L)$, it is interesting to study its local analogue,

$$D_l(L) = < (k - < k >_{L_0})^2 >_{L_0}, \tag{4.65}$$

where L_0 is the interval of local averaging and l is the coordinate of the left-hand border of this interval. An existence of a trend in the dependence $D_l(L)$ on l would be clearly indicative of non-stationarity of the stochastic process being studied. To verify the stationarity of the coarse-grained text of the Bible we perform the numerical simulation of the $D_l(L)$ dependence on l at different fixed values of L. As an example, the result of this simulation for $L = 10$ is shown in Fig. 19. It is clearly seen that there are regular fluctuations without a pronounced trend. The fluctuations result from the finiteness of the interval L_0 over which averaging is done. This fact allows us to make a conclusion about stationarity of the coarse-grained text of the Bible. The similar analysis of many other texts yielded the same result. It is expedient to study the global characteristics $D(L)$, Eq. (1.8), of the sequence instead of the local one, $D_l(L)$.

Memory function. According to Eqs. (2.109) and (2.117), the memory function can be restored using the variance or the correlation function. The MF thus obtained for the coarse-grained text of the Bible at $r < 300$ is given in Fig. 20. At long distances, $r > 300$, the memory function can be nicely approximated by the power function $F(r) = 0.27r^{-1.1}$, which is shown by the solid line in the inset in Fig. 20. Note that the persistent part of the

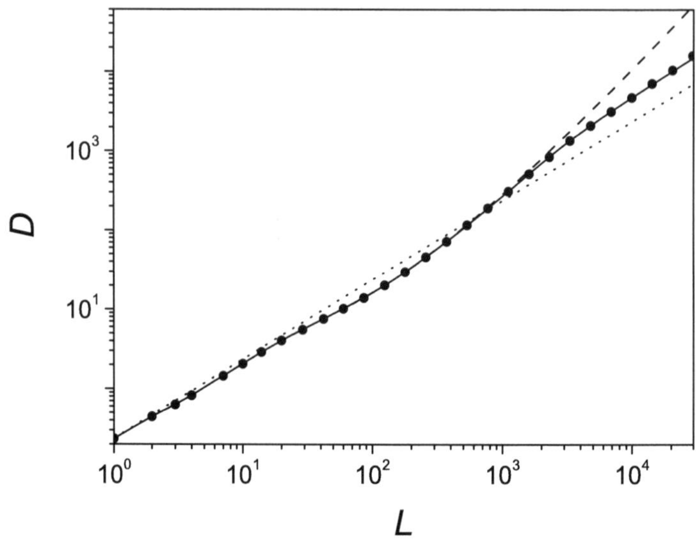

Figure 18 The variance $D(L)$ for the coarse-grained text ((letters with even numbers in the Alphabet) $\mapsto 1$, (ones with odd numbers) $\mapsto 0$) of the Bible (dashed line) and for the sequence obtained by shuffling the blocs of the length $L_0 = 3000$ (filled circles). The solid line represents the analytical results obtained from Eqs. (1.11) and (4.64). The dotted straight line describes the non-correlated Brownian diffusion, $D_0(L) = L\bar{a}(1-\bar{a})$.

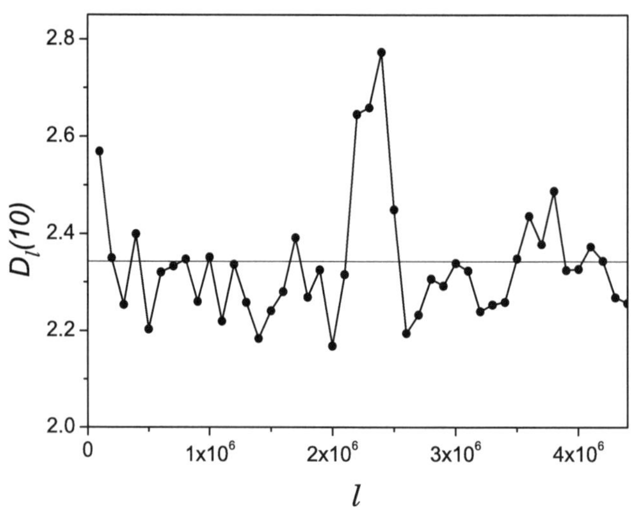

Figure 19 The local variance $D_l(10)$ for the coarse-grained text of the Bible vs the distance l. The averaging interval is $L_0 = 10^5$.

MF, $F(r > 300) \le 0.0008$, is much less than its typical magnitude 0.02 in the antipersistent region $r < 40$.

It should be emphasized that the short-range part of the memory function at $r \lesssim 40$ and the $D(L)$ function at $L \lesssim 300$ are largely dependent on the method of coarse-graining.

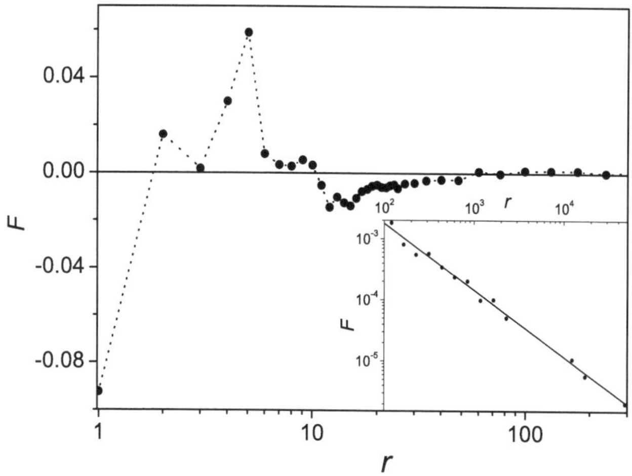

Figure 20 The memory function $F(r)$ for the coarse-grained text of the Bible at short distances. The power-law decreasing portion of the $F(r)$ plot for the Bible is represented by the filled circles in the inset. The solid line is for the power-law fitting.

Nevertheless, the antipersistent correlations exist for practically all kinds of the coarse-graining procedure. An interesting feature is that the region $r \lesssim 40$ of negative antipersistent memory function provides far longer distances $L \sim 300$ of the antipersistent behavior of the variance $D(L)$.

To test the appropriateness of modeling the coarse-grained texts by the additive multi-step Markov chains, we have generated the sequence of symbols using conditional probability (1.2) with the reconstructed memory function $F(r)$ of the text of Bible. The variance $D(L)$ of this sequence is plotted by the filled circles in Fig. 17. The nice coincidence of this curve with the solid line describing the variance of the initial coarse-grained text of the Bible proves the robustness of our method of the MF reconstruction.

In order to prove the universal character in the power-law decrease of the memory function at long distances, we have made a comparison between the MFs of the coarse-grained texts for more than fifty different literary works. The texts are coarse-grained by mapping the letters from the first and second halves of the alphabet onto zero and unity, respectively. Subsequently, using Eq. 2.117, we first calculate the variances and then the memory functions. All the curves for the memory functions can be well fitted by the power-law functions $F(r) = cr^{-b}$. The results of the fitting for eight texts written or translated into Russian [136, 137] are shown in Fig. 4.5.3. The exponents in all curves vary over the interval between $b_{min} = 1.02$ for "War and Peace" by L. Tolstoy and $b_{max} = 1.56$ for the Koran. Thus, the constants c and b can be used for linguistic classification of different literary works. It is significant that the memory functions for the texts of the English- and Russian-worded Bible, as well as the texts of the Old and New Testaments are practically coincident.

The existence of two characteristic regions having different behavior with reference to the memory function and, accordingly, to the persistent and antipersistent portions in the $D(L)$ dependence appears to be a prominent feature of all texts in any language. Note that the antipersistent portion of the memory function corresponds to the region where the grammatical rules are in use. Therefore, we call these kinds of correlations

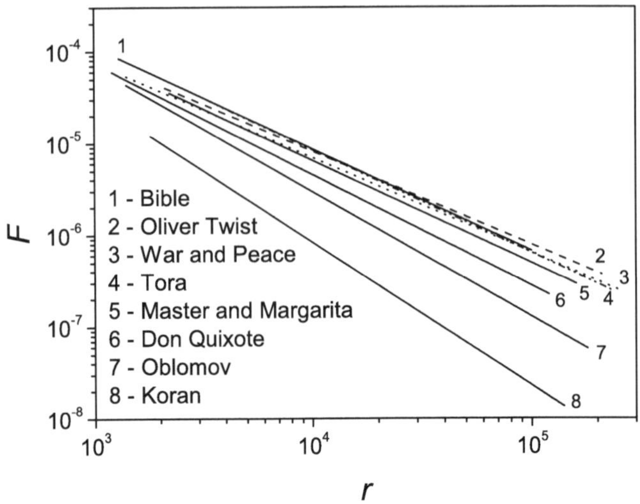

Figure 21 The memory function at long distances for the coarse-grained texts of eight literary works: 1. The Bible ($4.2 \cdot 10^6$ characters), 2. "Oliver Twist" by Charles Dickens ($8.9 \cdot 10^5$ characters), 3. "War and Peace" by Leo Tolstoy ($3.0 \cdot 10^6$ characters), 4. The Tora ($6.7 \cdot 10^5$ characters), 5. "Master and Margarita" by Mikhail Bulgakov ($8.0 \cdot 10^5$ characters), 6. "Don Quixote" by Miguel de Servantes ($2.0 \cdot 10^6$ characters), 7. "Oblomov" by Ivan Goncharov ($9.8 \cdot 10^5$ characters), 8. The Koran ($8.5 \cdot 10^5$ characters).

the "grammatical" ones. The persistent correlations in a text at very long distances can be related to a general idea of the literary work. Thus, these kinds of correlations are referred to as the "semantic" ones.

Two fundamentally different portions in the MF plots result from a peculiar competition between the two above-mentioned kinds of correlations. We would like to stress that both portions of the MF are equally important to provide an insight into the correlation properties of the literary texts. To support this statement we generate two special sequences. In both of them, only one kind of the memory function for the coarse-grained text of the Bible is taken into account, and the memory function in the other region is assumed to be zero. The variance $D(L)$ for these two sequences is given in Fig. 22. The lower (dashed) line represents the case where only the negative antipersistent portion, $r < 40$, of the memory function is allowed for. The upper (dash-dot-dotted) curve is for the sequence that is generated by means of the long-range persistent memory, $F(r) = 0.27r^{-1.1}$, $r > 100$. It is evident that the generated sequence with the antipersistent memory function displays the sub-diffusion only, whereas the sequence that corresponds to the persistent memory function is characterized by the super-diffusion behavior of the variance $D(L)$. The difference between the variances for two generated sequences and for the original coarse-grained text of the Bible, shown by the solid line in the same figure, corroborates our assumption about the significance of both kinds of the memory function.

Self-similarity of the coarse-grained texts. The power-law decrease (without characteristic scale) of the memory function at long distances leads to quite an essential property of *self-similarity* of the coarse-grained texts with respect to the *decimation* procedure discussed in Reference [32]. This procedure implies the deterministic or random removal of

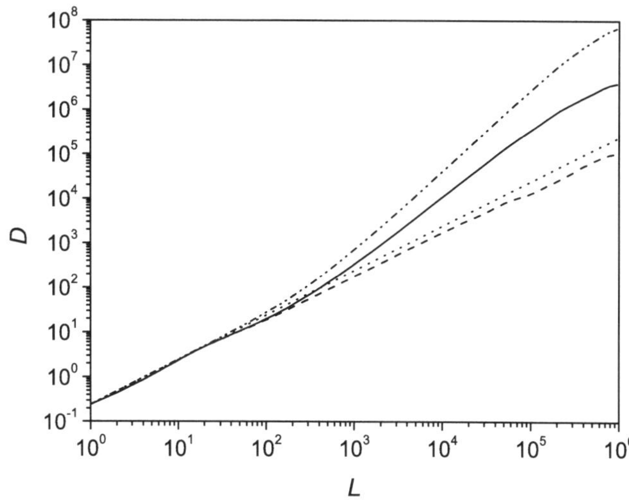

Figure 22 The variance $D(L)$ for the coarse-grained text of the Bible [135] (the solid line), and for the sequences constructed with using the persistent part of the MF (dash-dot-dotted line) and the antipersistent one (dashed line). The dotted line describes the non-correlated Brownian diffusion, $D_0(L) = L\bar{a}(1-\bar{a})$.

some part of symbols from a sequence and is characterized by the decimation parameter $\lambda < 1$ which represents the fraction of symbols kept in the chain. For example, under the random decimation each symbol is eliminated with probability $1-\lambda$. It can be shown that both of these procedures, deterministic and stochastic, are equivalent for a Markov chain. The sequence is self-similar if its variance $D(L)$ does not change after the decimation up to a definite value of L (which is dependent on the memory length of the original sequence and the decimation parameter). The model of the additive binary multi-step Markov chain with the step-wise MF (which was discussed in Reference [32]) offers the exact property of self-similarity at a length shorter than the memory length N. The coarse-grained literary texts are self-similar objects as well. It is indicated in Fig. 23 where three $D(L)$ curves are plotted for different values of the parameters of the regular decimation. Note that the decimation procedure leads to a decrease in the effective memory length. As a result, the variance curves coincide up to the effective memory depth, which is proportional to the decimation parameter. A similar phenomenon occurs in the case of random decimation as well.

A question arises: what particular property of the memory function is crucial for the self-similarity of the coarse-grained literary texts? It is natural to assume that the persistent long-range scale-free portion of the memory function affords this property because the self-similarity is specifically manifest at long distances. To verify this supposition we carry out the decimation procedure with different λ for the Markov chain constructed by using the long-range part of the Bible memory function only and then plot the corresponding $D(L)$ dependence. The curves are shown in the inset in Fig. 23. The self-similarity for this sequence appears to be more pronounced than for the original coarse-grained text of the Bible. Moreover, the antipersistent part of the MF disappears very fast after the decimation. This is clearly revealed as a disappearance of the antipersistent sub-linear portion of the

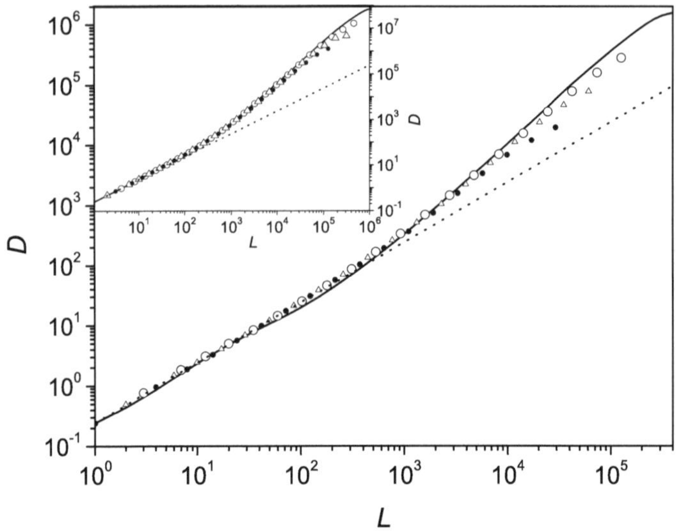

Figure 23 The numerically calculated variance $D(L)$ for the coarse-grained text of the Bible [135] (solid line) and for the sequences obtained after their regular decimation. The circles, triangles, and dots are for the decimation parameters 2, 4, and 8, respectively. The dotted line describes the non-correlated Brownian diffusion, $D_0(L) = L\bar{a}(1-\bar{a})$. The similar curves obtained for the sequence constructed by using the long-range part of Bible's memory function only are shown in the inset.

$D(L)$ curves in Fig. 23. After the decimation, the solid line transforms into the entirely persistent super-linear curve, which goes above the curve $D_0 = L\bar{a}(1-\bar{a})$. The conclusion about the invariance of the statistical properties of studied sequence with respect to the decimation procedure is an additional argument in favor of coarse-graining efficiency. The decimation can be thought of as additional coarse-graining of the initial random sequence.

It should be noted that there exist special kinds of short-range correlated texts which can be specified by both of the parameters, N and μ. For example, all dictionaries consist of the families of words where some preferred letters are repeated more frequently than in their other parts. Yet another example of the shortly correlated texts is any lexicographically ordered list of words. The analysis of written texts of this kind is given below.

Dictionaries. As an example, we have investigated the statistical properties of the coarse-grained alphabetical (lexicographically ordered) list of the most frequently used 15462 English words. In contrast to other texts, the statistical properties of the coarse-grained dictionaries are very sensitive to the kind of mapping. If we use the above-mentioned mapping, (a-m) $\to$ 0; (n-z) $\to$ 1, the behavior of the variance $D(L)$ similar to that shown in Fig. 15 would be obtained. The particular construction of the dictionary manifests itself if the preferred letters in the neighboring families of words are mapped onto the different symbols. The variance $D(L)$ for the dictionary coarse-grained by means of such mapping is shown by circles in Fig. 24. The graph of the function $D(L)$ consists of two portions similarly to the curve, which can be obtained for the generated N-step Markov sequence. The fitting of the numerical result in Fig. 24 by analytical itself (the curved solid line in Fig. 24) yields the values of the parameters $N = 180$ and $\mu = 0.44$.

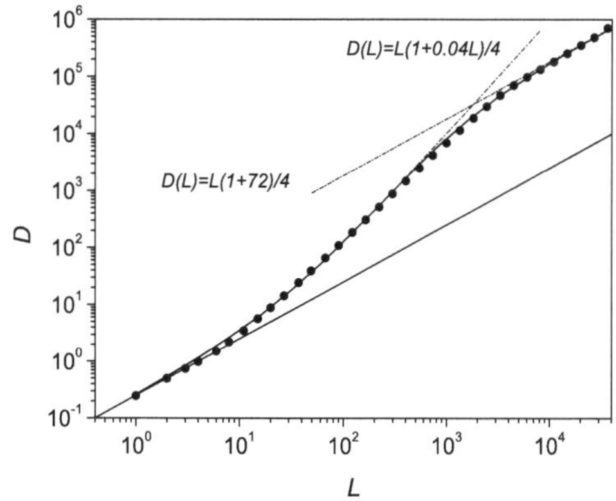

Figure 24 The dependence $D(L)$ for the coarse-grained alphabetical list of 15462 English words (circles). The curved solid line is the plot of function Eq. (55) with the fitting parameters $N = 180$ and $\mu = 0.44$. The straight solid line describes the non-correlated Brownian diffusion, $D(L) = L/4$.

4.5.4 DNA "texts"

In this Subsection the evolved theory of binary additive Markov chains with step-wise memory function (2.9) is applied to the coarse-grained DNA strings, [58]. It is known that any DNA text is written by four "characters", specifically by adenine (A), cytosine (C), guanine (G), and thymine (T). Therefore, there are three nonequivalent types of DNA text mapping onto one-dimensional binary sequences of zeros and unities. The first of them is the so-called purine-pyrimidine rule, {A,G} → 0, {C,T} → 1. The second one is the hydrogen-bond rule, {A,T} → 0, {C,G} → 1. And, finally, the third is {A,C} → 0, {G,T} → 1.

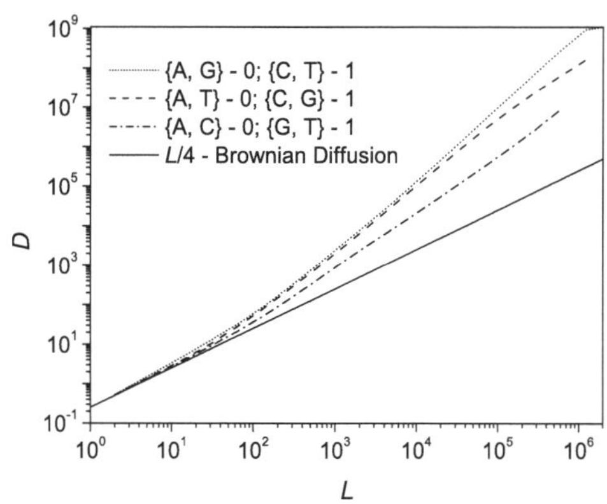

Figure 25 The dependence $D(L)$ for the coarse-grained DNA text of *Bacillus subtilis, complete genome*, for three nonequivalent kinds of mapping. Dotted, dashed, and dash-dotted lines correspond to the mappings {A,G} → 0, {C,T} → 1 (the parameter $m = 4.1 \cdot 10^{-2}$), {A,T} → 0, {C,G} → 1 ($m = 2.5 \cdot 10^{-2}$), and {A,C} → 0, {G,T} → 1 ($m = 1.5 \cdot 10^{-2}$), respectively. The solid line describes the non-correlated Brownian diffusion, $D(L) = L/4$.

By way of example, the variance $D(L)$ for the coarse-grained text of *Bacillus subtilis, complete genome* [63] is displayed in Fig. 25 for all possible types of mapping. We can see that the persistent properties of DNA are more pronounced than for the written texts and, contrary to the written texts, the $D(L)$ dependence for DNA does not exhibit the anti-persistent behavior at small values of L. In our view, the noticeable deviation of different curves in Fig. 25 from each other demonstrates that the DNA texts are much more complex objects in comparison with the written ones. Indeed, the different kinds of mapping reveal and emphasize various types of physical attractive correlations between the nucleotides in DNA, such as the strong purine-purine and pyrimidine-pyrimidine persistent correlations (the upper curve), and the correlations caused by a weaker attraction A↔T and C↔G (the middle curve).

It is interesting to compare the correlation properties of the DNA texts for three different species that belong to the major domains of living organisms: the Bacteria, the Archaea, and the Eukarya [63]. Figure 26 shows the variance $D(L)$ for the coarse-grained DNA texts of *Bacillus subtilis* (the Bacteria), *Methanosarcina acetivoransthe* (the Archaea), and *Drosophila melanogaster* - fruit fly - (the Eukarya) for the most representative mapping $\{A,G\} \rightarrow 0$, $\{C,T\} \rightarrow 1$. It is seen that the $D(L)$ curve for the DNA text of *Bacillus subtilis* is characterized by the highest persistence. As well as for the written texts, the $D(L)$ curves for the DNA of both the Bacteria and the Archaea do not contain the linear portions given by Ref. [58, Eq. (73)]. This suggests that their DNA chains are not stationary sequences. In this connection, we would like to point out that their DNA molecules are circular and represent the collection of extended coding regions interrupted by small non-coding regions. According to Figs. 25, 26, the non-coding regions do not disrupt the correlation between the coding areas, and the DNA systems of the Bacteria and the Archaea are fully correlated throughout their entire lengths. Contrary to them, the DNA molecules of the Eukarya have the linear structure and contain long non-coding portions. As evident from Fig. 26, the DNA sequence of the representative of the Eukarya is not entirely correlated. The $D(L)$ curve for the X-chromosome of the fruit fly corresponds qualitatively to [58, Eqs. (71) and (72)] with $\mu \approx 0.35$ and $N \approx 250$. If we draw an analogy between the DNA sequences and the literary texts, the resemblance of the correlation properties of integral literary novels and the DNA texts of the Bacteria and Archaea are to be found, whereas the DNA texts of the Eukarya are more similar to the collections of 10^4–10^5 short stories.

If our understanding as to the nature of correlations in the literary texts and DNA strings is correct, then several important questions would be of great interest, for example:

- Does the lack of the antipersistent portion in the memory function (and in the $D(L)$ dependence) in the DNA texts mean that the "grammatical rules" are non-existent in the "DNA language"?
- If we consider the variance $D(L)$ as a measure of information redundancy, can we explain the equality $D(L)_{|DNA} \simeq 10 \cdot D(L)_{|Text}$ resulting from the comparison between literary and DNA texts at $L \sim 3 \times 10^5$ in the following way: Nature is more careful about the conservation of the information stored in the DNA sequences than the Writer in his literary works?

4.5.5 Conclusion

Thus, we have demonstrated that the description of the literary works is suitable in terms of the Markov chains with complex memory functions. Actually, the memory function

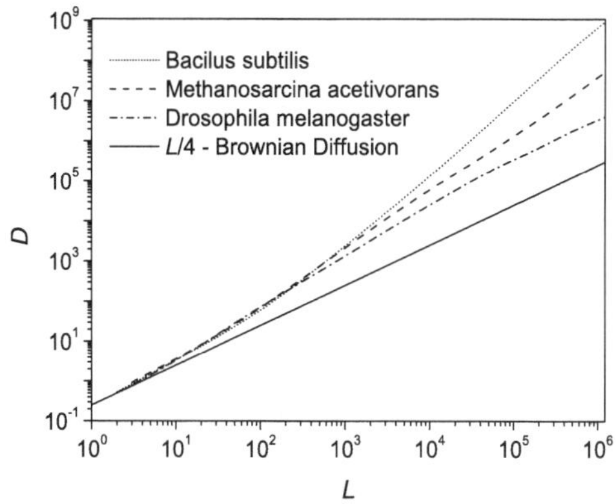

Figure 26 The dependence $D(L)$ for the coarse-grained DNA texts of *Bacillus subtilis, complete genome*, the Bacteria, (dotted line); *Methanosarcina acetivorans, complete genome*, the Archaea, (dashed line); *Drosophila melanogaster chromosome X, complete sequence*, the Eukarya, (dash-dotted line) for the mapping {A,G} $\to$ 0, {C,T} $\to$ 1. The solid line describes the non-correlated Brownian diffusion, $D(L) = L/4$.

appears to be a convenient informative "visiting card" of any symbolic stochastic process. We have studied the coarse-grained literary texts and shown the complexity of their organization in contrast to a previously discussed simple power-law decrease of correlations. We have proved that the competition between the two kinds of correlations govern the statistical properties of the coarse-grained texts. The antipersistent correlations exist at short distances, $L \lesssim 300$, in the region where the grammatical rules are efficient. The other kind of correlations, the persistent one, plays the main role at long distances, $L \gtrsim 300$. It can be related to the general idea of a literary work. Therefore, the first kind of correlations may be referred to as the grammatical one, whereas the second kind may be called as semantic correlations. The additional investigations are necessary to clarify the nature of the correlations. For example, it would be of interest to elaborate the dynamical model, e.g. a cellular automaton that produces the sequence with the same enigmatic long-range behavior of the memory function as the one in the literary works.

We have studied the simplest examples of random sequences, the dichotomic one. However, our preliminary considerations presented in the next Section suggest that the formulated concept of the additive Markov chains can by generalized to a larger class of random Markov processes with the finite or infinite number of states in the discrete or continuous "time". The suggested approach can be used for the analysis of other correlated systems in different fields of science.

4.6 SYMBOLIC DYNAMICS

An increasing interest in the theory of symbolic dynamics is stimulated by the need for a simple tool to investigate complicated dynamical systems, such as those considered in this Section. The symbolic dynamics have many features common to the numeric Markov chains with a finite state space. For this reason we present a short overview of the field of symbolic dynamics, give some elementary results and discuss coarse-graining procedure (partition). The readers can consult an excellent book on symbolic dynamics [138] to obtain more profound and detailed results.

The theoretical concepts of the symbolic analysis provide a unified framework for the acquisition of information about the flow of a dynamical system. The symbolic dynamics appeared as a tool for analyzing general dynamical systems by discretizing space. Specifically, computer simulations of continuous systems necessarily involve a discretization. In this way an "equivalent" system can be studied using symbolic dynamics thus representing trajectories as infinite length sequences (we look at the system infinitely long) using a finite number of symbols (a finite state space). Imagine a continuous dynamical system with representative point following some trajectory in a space. Let us divide the space into finitely many pieces, each labeled by a different symbol from some alphabet $\mathcal{A}$. We obtain a symbolic trajectory by writing down the sequence of symbols corresponding to the successive partition elements visited by the point in its orbit. An example of this idea is sequences of digits of real numbers. Can we learn anything about the dynamics of the system by scrutinizing its symbolic trajectories?

The answers to these questions will depend not only on the nature of our dynamical system, but also on the judicious choice of a partition (discretization).

According to many authors [140, 141, 139], the first successful use of symbolic dynamics techniques is generally attributed to J. Hadamard. In paper [142] the author constructed surfaces in $\mathbb{R}^3$ of negative curvature and discovered that geodesics on these surfaces can be described by sequences of symbols via a certain "coding" procedure. Forty years later the subject was under its first consistent study in the foundational paper of M. Morse and G. Hedlund [143]. Therein the symbolic systems are originally treated in the abstract form as objects in their own right. This study was motivated both by the intrinsic mathematical interest of symbolic systems and the need to get a better understanding of them in order to apply symbolic techniques to continuous systems. However, a further impulse was given by the emergence of the information theory and the mathematical theory of communication pioneered by C. E. Shannon [144]. Symbolic dynamics has continued to find applications to an ever-widening array of continuous systems: hyperbolic systems, maps of the interval, billiards, complex dynamics and so on. At the same time it contributes to problems arising in the storage and transmission of data. The results of symbolic dynamics help us understand how well, or how badly, the simulation may mimic the original.

The aim of this section is to consider simple mathematical models of dynamical systems, which, after introducing discretization (more frequently used physical synonym is coarse-graining and mathematical term is partition), allows an analytical treatment. Our main interest is to study the behavior of orbits x, $f(x)$, $f^2 = f(f(x))$, ..., $f^n(x)$, ..., where $f^n(x)$ denote the nth iteration, $f(f^{n-1}(x))$.

One-dimensional fractional transformations. Consider the unit interval $I = [0, 1)$ and the map f that sends $x \in I$ to $\{2x\}$, the fractional part of $2x$, Ref. [139]. We are interested in the orbit $x_0 = x$, $x_1 = f(x_0)$, $x_2 = f(x_1)$, ... We do not want to trace this orbit exactly and will limit ourselves with discretization of space: we will split I into just two parts, $I_0 = [0, 1/2)$ and $I_1 = [1/2, 1)$. We assign to x a symbolic trajectory $s_0 s_1 s_2 ...$, where s_i is 0 or 1 according as $f^i(x)$ is in I_0 or I_1. The expression $.s_0 s_1 s_2 ...$ is a binary expansion of the number x. Hence x is completely determined by its symbolic trajectory and vice versa. The map f now has a very interesting symbolic representation. If $x = .s_0 s_1 s_2 ...$ then $f(x) = \{2x\} = .s_1 s_2 s_3 ...$. The mapping $\{2x\}$ shifts the symbolic sequence to the left and discards the initial symbol. The utility of symbolic dynamics is that the dynamics is given by a simple coordinate shift. Dynamic properties that might have seemed elusive in the

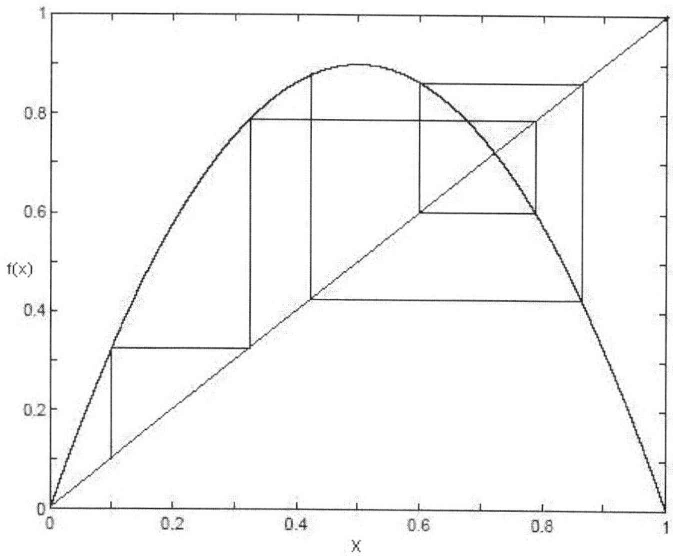

Figure 27 The unimodal quadratic map with initial point $x_0 = 0.1$.

original setting now become transparent. For instance, we can immediately identify the points of period 2 (that is, with $f^2(x) = x$). They are the two points with repeating symbolic trajectories $s_0 s_1 s_0 s_1 \ldots$, $f(x_0) = x_1$, $f(x_1) = x_0$.

Let us note that in the context of symbolical dynamics the iteration of the Gaussian map given by

$$S(x) = \left\{ \frac{1}{x} \right\} \tag{4.66}$$

leads to a rather unexpected expansion of irrational number $0 < \alpha < 1$ in a regular continued fraction

$$\alpha = \cfrac{1}{n_1 + \cfrac{1}{n_2 + \cfrac{1}{n_3 + \frac{1}{\cdots}}}}. \tag{4.67}$$

One of the fields of the chaotic dynamics, ergodic and metric theories is the studying of the distribution of sequences $\{n_i\}$ (see Reference 145).

One-dimensional unimodal transformations. Perhaps the simplest "dynamical system" and at the same time the prototypical example of a unimodal map $x_n \to x_{n+1} = f(x_n)$ that generates a symbolic sequence is the quadratic (logistic) map

$$x_{n+1} = \lambda x_n (1 - x_n). \tag{4.68}$$

The key topological property of a unimodal map is that it stretches and folds the interval $I = [0, 1]$ into itself. Figure (27) shows a trajectory of an initial point x_0 after six iterations. The iteration of the map is shown graphically by the plot of $y = f(x)$. For the input value $x = x_0$ the output is the value y, i.e. moving vertically to the graph of the function to $f(x)$.

To get the next input value x_1 we need the value $x=y$ given by moving horizontally to the diagonal. The next iteration x_2 is given by moving again vertically to $f(x)$, etc. In general, a unimodal map has a single critical point C in I and is the monotone increasing on the left of C and decreasing on the right. For the quadratic map, $I=[0, 1]$ and $C=1/2$.

To set up the symbolic dynamics of this system, we must first define a partition. Here, a partition is the separation of phase space into disjoint regions. For the quadratic map with $0<\lambda<4$, we define $I_0=[0, 1/2)$, $C=1/2$ and $I_1=(1/2, 1]$. Given x_0, we define x_n by Eq. (4.68) and

$$s_n = \begin{cases} 0, & x_n \in I_0, \\ C, & x_n = C, \\ 1, & x_n \in I_1. \end{cases} \tag{4.69}$$

The symbolic sequence or itinerary of point x_0 is the sequence $S(x_0)=\{s_0, s_1, s_2, ...\}$. As shown in previous example, the quadratic map acts on the sequence as a shift:

$$S(f^t(x)) = \sigma^t S(x) = \{s_t, s_{t+1}, s_{t+2}, ...\}, \tag{4.70}$$

where σ operates on sequences by discarding the left-most symbol, and shifting the rest of the sequence to the left. Note, in the previous example, for the sake of simplicity we omitted in partition the point $1/2$. It is common practice in mathematics to sacrifice the set of zero measure for the sake of simplicity or for other reasons.

One-dimensional Markov partition. Rigorous definition of the Markov transformation in $\mathbb{R}^1$ is as follows, Reference [146].

Let $I=[c, d]$ and let τ be the map of interval I onto itself: $I \to I$. Let $\mathcal{P}$ be a partition of I given by the points $c=c_0<c_1<...<c_p=d$. For $i=1, ..., p$, let $I_i=(c_{i-1}, c_i)$ be disjoint open intervals and denote the restriction of τ to I_i by τ_i. If τ_i is a homeomorphism (monotone and continuous mapping) from I_i *onto* a union of intervals of $\mathcal{P}$, then τ is said to be Markov's. The partition $\mathcal{P}$ is said to be the Markov partition with respect to the function τ. The illustration of the definition is shown in Fig. (28).

When the requirement is fulfilled, we can study the dynamical system as symbolic and the symbolic trajectory completely determine the orbit. In the case of the piecewise linear mapping the statistical properties of the system can be fully described in terms of the (one step) Markov chain, Ref. [147].

The Bernoulli shift and baker's transformation. Suppose $P=\{p_0, ..., p_{k-1}\}$ is a probability vector. It defines a probability on the finite set $\{0, ..., k-1\}$ by letting the probability of symbol i be p_i. Define a new probability space to be the product space $\{0, ..., k-1\}^{\mathbb{Z}}$ of all bi-infinite sequences of elements from a set of k symbols. Suppose that the probability of events i and j, $i \neq j$ is the product of p_i and p_j (events i and j are uncorrelated). The Bernoulli shift defined using Eq. (4.70) on the sequence from $\{0, ..., k-1\}^{\mathbb{Z}}$ by the probability vector P is denoted $B(P)$. The Bernoulli shift $B(1/2, 1/2)$ is thought of as flipping a fair coin.

The baker's transformation of the half-open unit square $[0, 1)^2$ is defined by

$$T(x, y) = \begin{cases} (2x, \frac{1}{2}y) & \texttt{mod 1 if } x<1/2 \\ (2x-1, \frac{1}{2}(y+1)) & \texttt{mod 1 if } x \geq 1/2. \end{cases} \tag{4.71}$$

For the baker's transformation, let symbols 0 and 1 be used for coding the left and right halves of the square $[0, 1/2) \times [0, 1)$ accordingly. This coding map is easily seen

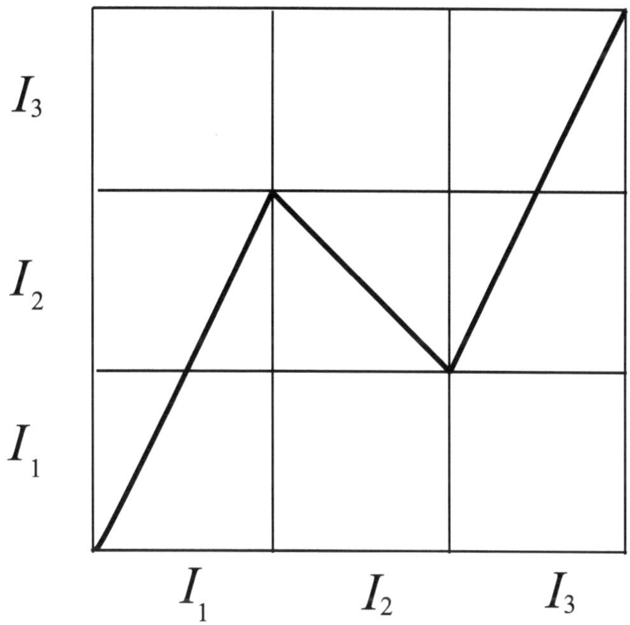

Figure 28 An example of a piecewise linear Markov map with the associated partition $\{I_1, I_2, I_3\}$.

to be a bijection between the baker's transformation and the Bernoulli shift $B(1/2, 1/2)$, Ref. [123]. The main idea of the proof is the presentation of the bi-infinite sequence $\{0, 1\}^{\mathbb{Z}} = \ldots, s_{-1}, s_0, s_1, \ldots$ by the point (x, y) on the torus:

$$x = \sum_{i=0}^{\infty} \frac{s_{-i}}{2^{i+1}}, \qquad y = \sum_{i=0}^{\infty} \frac{s_i}{2^i}. \tag{4.72}$$

Two-dimensional example - toral automorphism. The Arnold cat map, a uniformly hyperbolic (no eigenvalues on the unit circle) dynamical system is defined by

$$x = (Ax) \texttt{ mod } 1, \qquad A = \begin{pmatrix} 2 & 1 \\ 1 & 1 \end{pmatrix}. \tag{4.73}$$

This yields a well-defined map on the 2-torus from the unit square onto itself, which is invertible because $\texttt{Det}A = 1$, so A^{-1} also has integer entries. The map is said to be on the toral space T^2 because the "mod 1" operation causes the coordinate $1 + z$ to be equivalent to z. Matrix A has eigenvalues $0 < \lambda^{-1} < 1 < \lambda = (3 + \sqrt{5})/2$, and the corresponding eigenspaces provide the expanding and contracting directions at every point of T^2. Having an irrational slope, these spaces project to dense curves on T^2. It is also easy to check that the set of rational points is precisely the set of periodic points. The Arnold cat map is shown in Fig. (29). The cat map is the part of a larger class of functions called toral Anosov diffeomorphisms [145].

Symbolic Markov partitions in two dimensions. For a hyperbolic map A, symbolic dynamics relates the orbits of A to symbolic sequences by means of a partition of phase space, [148]. Such partition consists of a set of regions (cells) C_i, $i = 1, 2, \ldots, m$, covering

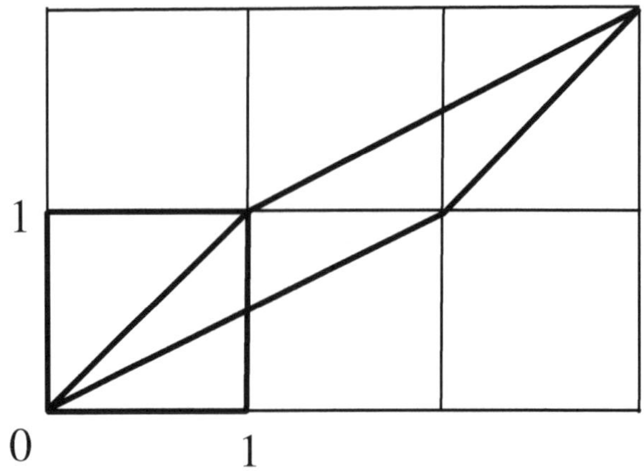

Figure 29 The parallelepiped is the image of the unit square for the cat map. After the "mod 1" operation, the image is exactly coincide with the unit square.

entirely the trajectory of system and then assigns symbols s_i to cells C_i, respectively. The regions are usually referred to as rectangles because their boundaries consist of pieces of stable and unstable manifolds (lines, surfaces). The Markov condition can be realized in a simple geometric way and has to satisfy the following properties. This is a suitable finite collection of closed subsets, whose interiors are nonempty, pairwise disjoint and whose union is the whole torus. Specifically, when projected to the unstable direction, the image of a rectangle goes entirely across every rectangle it meets, and when projected to the stable direction the preimage of a rectangle goes entirely across every rectangle it meets [145]. Such partitions are called the Markov partitions.

If the above-mentioned conditions are not fulfilled, there is no one-to-one correspondence between points in the phase space and those in the symbolic space. Such an ill-defined symbolic representation of phase-space points is not desirable for theoretical analysis and applications. Symbolic representations of controlled chaotic orbits produced by signal generators can be used for communication [149].

For the details about the construction of a Markov partition reflecting the geometrical action of a hyperbolic automorphism of the n-torus see Reference [150].

Finite dynamical systems. The term *finite dynamical systems* is frequently used in the wide sense as a synonym of "symbolical systems". In the narrow sense of the word it is one branch of symbolic dynamics, i.e., "finite dynamical system" which implies a mapping from the set of binary strings of a certain length onto itself. Such mappings occur in a variety of contexts, in particular in the theoretical study of computer simulations. Representing a computer simulation as a particular finite dynamical system is one possible approach to its mathematical analysis, [151].

So, we presented some examples of the transitions from the continuous-valued systems to symbolic systems, the so-called coarse graining. It should be emphasized that this problem in question is too wide to be completely and comprehensively reviewed here.

5. Conclusions

The examined model of binary additive multi-step Markov chains deals with the simplest case of numerical chains that can reproduce a given correlation properties of random systems. However, this model can be generalized and extended to some more complex mathematical objects. In this Section we describe some ideas of further developments and generalizations of the additive Markov chains model.

n-valued sequence and the symbolic correlator. One of the ways to generalize the additive model is to consider the *non-binary* Markov chain a_i, consisting of the **symbols** 1, ..., n. Such systems are usually characterized by a more complex property than the binary correlator, namely by the *symbolic correlation function* which can be defined as

$$K_{\alpha\beta}(r) = \langle [\delta(a_i - \alpha) - \bar{\alpha}][\delta(a_{i+r} - \beta) - \bar{\beta}] \rangle. \tag{5.1}$$

Here $\delta(...)$ denotes the Kroneker delta-symbol,

$$\bar{\alpha} = \langle \delta(a_i - \alpha) \rangle,$$

the angle brackets means the averaging over index i.

Function $K_{\alpha\beta}(r)$ possesses the properties of correlation function: if the value of $K_{\alpha\beta}(r)$ equals zero, then the events $(a_i = \alpha)$ and $(a_{i+r} = \beta)$ are statistically independent. The positive values of K correspond to the persistent case, when the symbol $a_i = \alpha$ increases the probability of symbol $a_{i+r} = \beta$ occurring. The negative values of K correspond to the anti-persistent case.

It is easy to show that the symbolic correlation function defined in this way has the following properties:

$$\begin{gathered} K_{\alpha\beta}(0) = \bar{\alpha}\delta(\alpha - \beta) - \bar{\alpha}\bar{\beta};\ K_{\alpha\beta}(r) = K_{\beta\alpha}(-r); \\ \sum_{\alpha} K_{\alpha\beta}(r) = \sum_{\beta} K_{\alpha\beta}(r) = 0. \end{gathered} \tag{5.2}$$

Here the symbol $\sum_{\alpha}$ means the summation over all possible values of symbol α: 1, 2, ..., n.

n-valued N-step Markov chain. If the probability of some symbol occurring depends on N previous symbols only then such a sequence will be referred to as *n-valued N-step Markov chain*. The statistical properties of the sequence are determined by the *conditional probability function*. This function determines the probability of the symbol $a_i = \alpha$ occurring at the i-th site on condition that the values of all N **previous** symbols are fixed.

Then we suppose that the considered Markov chain is *additive*, i.e. the conditional probability function has an additive form,

$$P(a_i = \alpha | \dots) = \bar{\alpha} + \sum_{r} \sum_{\beta} F_{\beta\alpha}(r)(\delta(a_{i-r} - \beta) - \bar{\beta}). \tag{5.3}$$

It means that the previous symbols exert an independent influence on the probability of the definite symbol occurring.

There is no one-to-one relation between the memory function $F_{\beta\alpha}(r)$ and the conditional probability function $P(a_i = \alpha)$. Indeed, the function $F_{\beta\alpha}(r) = F'_{\beta\alpha}(r) + \varphi_\alpha(r)$ provides the same probability as $F'_{\beta\alpha}(r)$.

The additional term $\varphi_\alpha(r)$ can be selected for the reason of convenience. Consider a sum

$$\sum_\beta \bar{\beta} F_{\beta\alpha}(r) = \sum_\beta \bar{\beta} F'_{\beta\alpha}(r) + \varphi_\alpha(r).$$

We can fix the value of $\varphi_\alpha(r)$ by imposing the following condition,

$$\sum_\beta \bar{\beta} F_{\beta\alpha}(r) = 0. \tag{5.4}$$

Then the conditional probability function takes on the form

$$P(a_i = \alpha | \ldots) = \bar{\alpha} + \sum_r \sum_\beta F_{\beta\alpha}(r) \delta(a_{i-r} - \beta). \tag{5.5}$$

Another condition for the memory function follows from the normalization of the probability $P(a_i = \alpha)$:

$$\sum_\alpha F_{\beta\alpha}(r) = 0. \tag{5.6}$$

Connection between the memory and the correlation functions. As with a binary case, we can write a connection between the memory $F_{kj}(r')$ and the correlation functions $K_{\alpha\beta}(r)$. For any values of $\alpha, \beta = 1, 2, ..., n$ and $r = 1, 2, ...$ we have the following equation

$$K_{\alpha\beta}(r) = \sum_{r'=1}^{N} \sum_\gamma K_{\alpha\gamma}(r - r') F_{\gamma\beta}(r'). \tag{5.7}$$

Symmetric n-valued chain. We can consider a more simple special case of n-valued chain. First, suppose that the statistical properties are equal for all symbols $\alpha = 1, 2, ..., n$. Second, the previous symbols $\beta \neq \alpha$ (located at some distance r) has an equal influence on the current symbol α, and the symbol $\beta = \alpha$ exhibits a different impact. In terms of the memory function it implies the following:

$$F_{\alpha\beta}(r) = \begin{cases} A(r), & \alpha = \beta \\ B(r), & \alpha \neq \beta; \end{cases} \qquad \bar{\alpha} = \frac{1}{n}.$$

Upon normalization, this equation yields

$$F_{\alpha\beta}(r) = F(r) \left(\delta_{\alpha\beta} - \frac{1}{n} \right). \tag{5.8}$$

It is easy to show that this memory function corresponds to the correlation function of the form,

$$K_{\alpha\beta}(r) = C(r)\left(\delta_{\alpha\beta} - \frac{1}{n}\right), \tag{5.9}$$

where $C(r)$ satisfies the set of equations,

$$C(r) = \sum_{r'=1}^{N} C(r - r')F(r'), \quad C(0) = \frac{1}{n}. \tag{5.10}$$

Numerical correlation function. In the above subsections, we considered a sequence of **symbols** a_i which can take on one of n values $1, 2, ..., n$. Suppose that every symbol a_α corresponds to definite numeric value ε_α. Then for **numeric** sequence $\varepsilon(i)$ an ordinary "numeric" correlation function can be studied:

$$K(r) = \langle (\varepsilon(i) - \bar{\varepsilon})(\varepsilon(i + r) - \bar{\varepsilon}) \rangle. \tag{5.11}$$

There is a simple relation between this quantity and the symbolic correlation function $K_{\alpha\beta}(r)$,

$$K(r) = \sum_{\alpha\beta} \varepsilon_\alpha \varepsilon_\beta K_{\alpha\beta}(r). \tag{5.12}$$

Multidimensional systems. The additive Markov chains can be associated with one-dimensional objects. As one of generalizations we can deal with multidimensional objects. For example, in 2D case the probability of definite symbol $a(x, y)$ occurring at discrete coordinates x and y depends on the symbols located in "left-top" rectangle $x' \in (x - N_x; x)$, $y' \in (y - N_y; y)$. For such 2D "Markov chains" we can represent a connection between the memory matrix-function $F(x, y)$ and correlation matrix-function $K(x, y)$. This model allows us to construct 2D objects with prescribed 2D-correlators.

Continuous time and continuous values. A binary 1D discrete sequence can be generalized to the continuous case. First, to model a time continuity, we should increase the memory length N of the binary chain and reduce the magnitude of memory function at the same value. Second, continuousness of the values can be obtained by increasing a number n in the n-valued chain model.

Non-stationary chains. This case is of great importance because there exist many physical systems with non-stationary behavior. It is possible to employ the time-dependent memory function $F(r, i)$,

$$P(a_i = 1|\ldots) = \bar{a} + \sum_r F(r, i)\delta(a_{i-r} - \beta), \tag{5.13}$$

and time-dependent correlation function $K(r_1, r_0)$. Therefore, we derive the following relation,

$$K(r_1, r_0) = \sum_r K(r_1, r_0 - r)F(r, r_0). \tag{5.14}$$

This relation can be solved as a non-stationary recurrence relation with initial condition $K(r_1, r_0) = K(r_0, r_1)$.

Modeling stochastic processes. The given stochastic process $h(t)$ can be modeled by another stochastic process derived from the binary Markov chain $a_i \in \{0, 1\}$. We suppose that it is normalized so that it takes on values from interval $(0, 1)$ and its derivative $h'(t)$ is limited. For example, the sums

$$k_i(L, M) = \frac{1}{L}\sum_{l=1}^{L} a_{iM+l}. \tag{5.15}$$

approximate the given stochastic process if parameters L and M are appropriately chosen. The correlation properties of the stochastic process $h(t)$ can be expressed in terms of the memory function of the binary additive Markov chain.

Quantum Markov chains. Let Ω be a set of different states (configurations) of a quantum system. We may introduce σ-algebra B of different quantum events, i.e. the subsets of Ω, and complex measure A on B such as $A(\Omega) = 1$ and $|A(\Lambda)| \leq 1$. The latter can be treated as the complex amplitude of an event, so that $|A(\Lambda)|^2$ is the probability of event Λ. Let $f : B \to Q$ be the measurement of some physical quantity, Q being the set of possible results of these measurements. Assume all the probabilities $P(f^{-1}(q)),\quad q \in Q$ to be non-zero. In this case we define measurement $g : B \to Q$ not interfering with f as the one that satisfies the following condition,

$$\sum_{q\in Q} P(g^{-1}(q)|f^{-1}(p)) = 1.$$

Note that the property of non-interfering is not commutative: the fact that g does not interfere with f does not imply that f does not interfere with g. N-chain is a sequence f_i, $i = 1..N$ of measurements with the same outcome space Q if $f_0^{-1}(q_0) = \Omega$ and all the probabilities of $f_t^{-1}(q_t) \bigcap f_{t-1}^{-1}(q_{t-1}) \bigcap \cdots \bigcap f_1^{-1}(q_1)$ are non-zero for all $t, q_1, \ldots, q_t$. The N-chain is stationary if

$$A(f_t^{-1}(q)|f_{t-1}^{-1}(p)) = A(f_2^{-1}(q)|f_1^{-1}(p)).$$

The N-chain is the Markov chain if

$$A(f_t^{-1}(q_t)|f_{t-1}^{-1}(q_{t-1}) \bigcap \cdots \bigcap f_1^{-1}(q_1)) = A(f_t^{-1}(q_t)|f_{t-1}^{-1}(q_{t-1})).$$

The N-chain is unitary if f_2 does not interfere with f_1 and

$$\sum_{r\in Q} A(f_2^{-1}(r)|f_1^{-1}(p))\bar{A}(f_2^{-1}(r)|f_1^{-1}(q)) = 0, \quad p \neq q.$$

Finally, the quantum Markov N-chain is the unitary stationary Markov N-chain.

It can be seen that $A(f_t^{-1}(q_t)|f_{t-1}^{-1}(q_{t-1}))$ is the complex amplitude of transition of the system from the state in which the measurement f_{t-1} gives the result q_{t-1} to the state in which the measurement f_{t-1} yields q_{t-1}. Thus f_i are the functions of a discrete stochastic process that generates these amplitudes. This means that the probabilistic evolution of the quantum system gives rise to the quantum Markov chain.

More generally (see Refs. [153, 152]), the quantum Markov chain is a pair (E, ρ) of quantum channel E and density matrix ρ such that

$$E: \quad B \otimes B \longrightarrow B \tag{5.16}$$

is a completely positive trace-preserving map, and B is a C*-algebra of bounded operators. The quantum Markov condition

$$Tr\rho(b_1 \otimes b_2) = Tr\rho E(b_1, b_2) \tag{5.17}$$

should be satisfied for all b_1 and b_2 from B.

One of the numerous applications of the quantum Markov chains is the modelling of signals created in the human brain [154]. The corresponding procedure may be thought of as one of choosing a signal from the set of signals contained in the memory. Using quantum statistics in lieu of classical allowed us to get a more correct insight into the nature of the signal generating. The methods discussed in this review can help and contribute to the study of these systems.

Appendix

A. TRANSITIVITY OF THE N-STEP MARKOV CHAINS

In this Appendix, we prove the property of metrical transitivity of the N-step Markov chains.

It is possible to look at the Markov chain from the other point of view and consider it as a 1-step *vector* Markov chain. To this end, we introduce the N-component vector-function X_l,

$$X_l = (a_{l+1}, a_{l+2}, ..., a_{l+N}), \qquad l = ..., -2, -1, 0, 1, 2, ... \tag{A.1}$$

The number of different sets of symbols $(a_{i+1}, a_{i+2}, ..., a_{i+N})$ is equal to $Q = 2^N$. We number the different states of the vector X_l by their binary representation,

$$D(a_N, a_{N-1}, ..., a_1) = a_N 2^0 + a_{N-1} 2^1 + ... + a_1 2^{N-1}, \qquad 0 \leq D \leq 2^N - 1. \tag{A.2}$$

The matrix elements M_{ik} of the probability matrix M, i.e. the probabilities of transition of the vector $X = (a_1, a_2, \ldots, a_N)$ into the vector $Y = (a'_1, a'_2, \ldots, a'_N)$ can be expressed via the function of conditional probability $P(a_i \mid T_{N,i})$. The subscripts i and k of the matrix M_{ik} are determined by the binary representations of the sequences $(a_1, a_2, \ldots, a_N)$ and $(a'_1, a'_2, \ldots, a'_N)$, correspondingly: $i = 1 + D(a_N, a_{N-1}, \ldots, a_1)$ and $k = 1 + D(a'_N, a'_{N-1}, \ldots, a'_1)$. Every matrix row contains only two non-zero elements since the vector X_1 can take up two values only, namely, $(a_2, a_3, \ldots, a_N, 0)$ and $(a_2, a_3, \ldots, a_N, 1)$. For $k \leq Q/2$, let us denote the probability of occurring of $a_{N+1} = 0$ as $1 - P_k$, where the index k is equal to $k = 1 + D(a_N, a_{N-1}, \ldots, a_1)$ in the binary representation.

For the index k being in the range from $Q/2 + 1$ to Q, we denote the probability of occurring of symbol $a_{N+1} = 0$ after the word $a_1, a_2, \ldots, a_N$ as P_k. Then, $1 - P_k$ is the probability of occurring of the symbol unity. Taking into account that $a_N = 0$ for $k \leq Q/2$ and obvious relations,

$$D(a_{N-1}, \ldots, a_1, 0) = 2D(a_N, a_{N-1}, \ldots, a_1),$$

$$D(a_{N-1}, \ldots, a_1, 1) = 2D(a_N, a_{N-1}, \ldots, a_1) + 1,$$

we get the transition probabilities matrix M:

$$M = \begin{pmatrix} 1-P_1 & P_1 & 0 & 0 & \ldots\ldots & 0 & 0 \\ 0 & 0 & 1-P_2 & P_2 & 0\ \ldots & 0 & 0 \\ \ldots & \ldots & \ldots & \ldots & \ldots\ldots & \ldots & \ldots \\ 0 & 0 & \ldots & \ldots & 0\ 0 & 1-P_{Q/2} & P_{Q/2} \\ 1-P_{Q/2+1} & P_{Q/2+1} & 0 & 0 & \ldots\ldots & 0 & 0 \\ 0 & 0 & 1-P_{Q/2+2} & P_{Q/2+2} & 0\ \ldots & 0 & 0 \\ \ldots & \ldots & \ldots & \ldots & \ldots\ldots & \ldots & \ldots \\ 0 & 0 & \ldots & \ldots & 0\ 0 & 1-P_Q & P_Q \end{pmatrix}. \tag{A.3}$$

Thus, to determine the vectors b of probability distribution of N-words in the stationary Markov chain we need to solve the system of equations,

$$b_i = \sum_{k=1}^{Q} b_k M_{ki}, \qquad \sum_{k=1}^{Q} b_k = 1. \tag{A.4}$$

In other words we need to obtain the normalized eigenvector corresponding to the eigenvalue $\lambda = 1$ of the matrix $M^{(N)}$ of the order $Q = 2^N$. It is clear that if the vector b satisfies to the condition $bM = b$ then for every integer k the condition $bM^k = b$ is also true, here M^k is the power k of the matrix M. Let us consider the matrix M^N and show that all matrix elements are positive. In this case, following the Markov theorem we can conclude that the matrix M determines uniquely the probability of the words distribution.

Let us suggest that for any $k < N$ the matrix M^k satisfies to the next conditions: in the first row the elements M_{1i} for $i = 1, \dots, 2^k$ are positive, in the second row the positive elements are M_{2i} with $i = 2^k + 1, \dots 2 \times 2^k$, ... in the 2^{N-k}-th row — $i = (2^{N-k} - 1)2^k, \dots, 2^N$. In the next rows this order is repeated. Let us demonstrate that if the matrix M^k obeys to this rules, then it is true for the matrix M^{k+1} also.

After multiplication of matrixes M^k and M the elements of obtained matrix are defined by the expression:

$$M^{k+1}(i, j) = \sum_{l} M^k(i, l) M(l, j). \tag{A.5}$$

Let us consider the first row of the matrix M^{k+1} — $i = 1$. In each column of the matrix M only two elements are non-zero. After multiplication of the first row of the matrix M^k to some column of the matrix M the result is non-zero (positive) for $j \leq 2 * 2^k$ only, because positive elements of the matrix M corresponds to the positive zone ($i < 2^k$) of the first row of matrix M^k only for this j. So the described rule remains for the first row of the matrix M^{K+1}. Similarly this fact can be proved for other rows.

The matrix M^1 obeys to this rule, consequently, by induction, it is true for all M^k. In according to this rule, if power $k = N$, then all elements of the matrix M^N are positive.

Therefore, from the Markov theorem, there is the unique solution of the system $bM^N = b$ (or $bM = b$). This solution can be obtained by the method of successive approximations,

$$b_i^{k+1} = b_j^k M(j, i), \qquad k = 0, 1, 2, \dots, \tag{A.6}$$

if we start from the arbitrary initial distribution b_j^0. In the limit $k \to \infty$ we get to the stationary distribution of the probability vector b.

Taking into account the explicit form of the matrix M, the equation (A.4) comes to the next equations:

$$b_i(1 - P_i) + b_{i+Q/2} P_{i+Q/2} = b_{2i-1}, \tag{A.7}$$

$$b_i P_i + b_{i+Q/2}(1 - P_{i+Q/2}) = b_{2i}.$$

For $Q = 2$ we get the well known result [155]:

$$M = \begin{pmatrix} 1 - P_1 & P_1 \\ P_2 & 1 - P_2 \end{pmatrix},$$

$$b_1 = \frac{P_2}{P_1 + P_2}, \qquad b_2 = \frac{P_1}{P_1 + P_2}.$$

In the case $Q = 4$ we obtain the following result:

$$M = \begin{pmatrix} 1-P_1 & P_1 & 0 & 0 \\ 0 & 0 & 1-P_2 & P_2 \\ P_3 & 1-P_3 & 0 & 0 \\ 0 & 0 & P_4 & 1-P_4 \end{pmatrix},$$

$$b_1 = \frac{P_3 P_4}{P_1 P_2 + 2P_1 P_4 + P_3 P_4}, \quad b_2 = b_3 = \frac{P_1 P_4}{P_1 P_2 + 2P_1 P_4 + P_3 P_4},$$

$$b_4 = \frac{P_1 P_2}{P_1 P_2 + P_1 P_4 + P_3 P_4}.$$

B. EQUATIONS FOR THE FOURIER TRANSFORMATIONS OF THE MEMORY AND CORRELATION FUNCTIONS

Equation for correlation function. We shall derive the Fourier transform of relation (2.119). First, we represent this equation in the following form,

$$\hat{\mathcal{K}}(k) = 1 + \sum_{r,r'=-\infty}^{\infty} \exp(ikr)\mathcal{K}(r-r')F(r') - \sum_{r=-\infty}^{\infty} \mathcal{K}(r)F(r)$$

$$-2\sum_{r,r'=1}^{\infty} \cos(kr)\mathcal{K}(r+r')F(r'). \tag{B.1}$$

Now we substitute the memory function $F(r')$ due to Eq. (2.126), together with the use of Eqs. (2.123). In this way we arrive at the relation,

$$\hat{\mathcal{K}}(k)\hat{F}(k) - 2\sum_{r,r'=1}^{\infty} \cos(kr)\mathcal{K}(r+r')F(r') = \hat{\mathcal{K}}(k) + A - 1, \tag{B.2}$$

where the constant A is defined by

$$A = \sum_{r=-\infty}^{\infty} \mathcal{K}(r)\, F(r) = \frac{1}{2\pi}\int_{-\pi}^{\pi} dk\, \hat{\mathcal{K}}(k)\, \hat{F}(k). \tag{B.3}$$

Into the second term of the left-hand side of Eq. (B.2) we substitute the correlator $\mathcal{K}(r+r')$ in the form of the Fourier integral (2.122). As a result, we get the integral equation B.4 for the power spectrum $\hat{\mathcal{K}}(k)$ of the binary correlator.

$$\left[\hat{F}(k) - 1\right]\hat{\mathcal{K}}(k) - \frac{1}{2\pi}\int_{-\pi}^{\pi} dk'\, \Upsilon(k,k')\hat{\mathcal{K}}(k') = A - 1. \tag{B.4}$$

Here the integral kernel $\Upsilon(k, k')$ is described as follows,

$$\Upsilon(k,k') = 2\sum_{r=1}^{\infty}\cos(kr)\exp(-ik'r)\sum_{r'=1}^{\infty}F(r')\exp(-ik'r') \tag{B.5}$$

$$= \frac{1}{2}\,Q(k,k')\frac{1}{\pi}\int_0^{\pi}dk''\hat{F}(k'')Q(k'',k'), \qquad k' = \lim_{\epsilon\to+0}(k'-i\epsilon),$$

and the Q-functions are defined below by Eq. (B.11).

Due to a complex nature of Q-functions, the kernel $\Upsilon(k,k')$ is also a complex function. Then, in accordance with the definition (B.5), we have,

$$\frac{1}{2\pi}\int_{-\pi}^{\pi}dk\,\Upsilon(k,k') = \frac{1}{\pi}\int_0^{\pi}dk\,\Upsilon(k,k') = 0, \tag{B.6}$$

$$\frac{1}{2\pi}\int_{-\pi}^{\pi}dk'\,\Upsilon(k,k') = \frac{1}{\pi}\int_0^{\pi}dk'\,\Upsilon(k,k') = 0. \tag{B.7}$$

When integrating Eq. (B.4) over the wave number k within the interval $(-\pi,\pi)$ or $(0,\pi)$, its integral term vanishes due to Eq.(B.6), while the equation itself turns into identity $A-1\equiv A-1$. This fact implies that the relation (B.3) automatically satisfies the solution of Eq. (B.4), independently on the value of A. Therefore, the constant A is determined solely by the normalization condition $\mathcal{K}(0)=1$ or, the same, by Eq. (2.125) for the power spectrum $\hat{\mathcal{K}}(k)$.

Equation for memory function. Let us now substitute the memory function (2.126) into the left-hand side of Eq. (B.2). In this way we obtain the integral equation for the Fourier transform $\hat{F}(k)$ of the memory function,

$$\hat{\mathcal{K}}(k)\hat{F}(k) - \frac{1}{\pi}\int_0^{\pi}dk'\,\Xi(k,k')\hat{F}(k') = \hat{\mathcal{K}}(k) + A - 1. \tag{B.8}$$

In this equation the kernel $\Xi(k,k')$ of the integral operator is described by the expressions

$$\Xi(k,k') = 2\sum_{r,r'=1}^{\infty}\cos(kr)\mathcal{K}(r+r')\cos(k'r') \tag{B.9}$$

$$= \frac{1}{4\pi}\int_{-\pi}^{\pi}dk''\hat{\mathcal{K}}(k'')Q(k,k'')Q(k',k''), \tag{B.10}$$

where the Q-functions are defined as

$$Q(k,k'') = -2\sum_{r=1}^{\infty}\cos(kr)\exp(-ik''r) \tag{B.11}$$

$$= 1 + i\frac{\sin k''}{\cos k - \cos k''}, \qquad k'' = \lim_{\epsilon\to+0}(k''-i\epsilon). \tag{B.12}$$

In spite of a complex nature of the Q-functions, the kernel $\Xi(k,k')$ is real, even and symmetrical function of both arguments k and k',

$$\Xi(-k,k') = \Xi(k,-k') = \Xi(k,k') = \Xi(k',k). \tag{B.13}$$

We can show that the mean value of the function $Q(k, k'')$ vanishes,

$$\frac{1}{2\pi}\int_{-\pi}^{\pi} dk\, Q(k, k'') = \frac{1}{\pi}\int_{0}^{\pi} dk\, Q(k, k'') = 0. \tag{B.14}$$

Indeed, integrating Eq.(B.12) over k, we can obtain,

$$\frac{1}{2k_c}\int_{-k_c}^{k_c} \frac{\sin k'' dk}{\cos k - \cos k''} = \frac{i\pi}{k_c}\Theta(k_c - |k''|) + 2\sum_{r=1}^{\infty}\frac{\sin(k_c r)}{k_c r}\sin(k'' r),$$

$$= \frac{i\pi}{k_c}\Theta(k_c - |k''|) + \frac{1}{k_c}\ln\left|\frac{\sin[(k_c + k'')/2]}{\sin[(k_c - k'')/2]}\right|; \quad 0 < k_c \leq \pi; \quad |k''| \leq \pi. \tag{B.15}$$

The first expression for the integral is obtained by a direct integration of the sum in Eq.(B.11). In the derivation of the second expression we have used the fact that the Heaviside unit-step function $\Theta(x)$, for which $\Theta(x<0)=0$ and $\Theta(x>0)=1$, arises due to the contribution of simple poles $k = \pm(k'' - i\epsilon)$, $\epsilon \to +0$. Clearly, this contribution emerges only when the poles are within the integration interval, i.e. if $|k''| < k_c$. The principal value (*P.V.*) of the integral in Eq.(B.15) has the following form,

$$P.V. \int_0^{k_c} \frac{\sin a\, dx}{\cos x - \cos a} = \frac{1}{2} P.V. \int_0^{k_c} dx\left[\cot\left(\frac{x+a}{2}\right) - \cot\left(\frac{x-a}{2}\right)\right]$$

$$= \ln\left|\frac{\sin[(k_c + a)/2]}{\sin[(k_c - a)/2]}\right|. \tag{B.16}$$

We can see that the integral (B.16) vanishes when $k_c = \pi$, and the Q-function property (B.14) is evident.

Note that Eq.(B.14) gives rise to the normalization relations for the kernel $\Xi(k, k')$,

$$\frac{1}{2\pi}\int_{-\pi}^{\pi} dk\, \Xi(k, k') = \frac{1}{\pi}\int_{0}^{\pi} dk\, \Xi(k, k') = 0, \tag{B.17}$$

$$\frac{1}{2\pi}\int_{-\pi}^{\pi} dk'\, \Xi(k, k') = \frac{1}{\pi}\int_{0}^{\pi} dk'\, \Xi(k, k') = 0. \tag{B.18}$$

As a result, the integration of Eq. (B.8) over the wave number k within the interval $(-\pi, \pi)$ or $(0, \pi)$ leads to the identity $A \equiv A$. This fact implies that the relation (B.3) automatically satisfies the solution of Eq. (B.8). Therefore, the constant A is determined by the initial condition for the memory function $F(r)$,

$$F(0) = \frac{1}{2\pi}\int_{-\pi}^{\pi} dk\, \hat{F}(k) = \frac{1}{\pi}\int_{0}^{\pi} dk\, \hat{F}(k) = 0. \tag{B.19}$$

C. DISTRIBUTION FUNCTION OF β_n

A standard way to derive the probability density $\rho_B(\beta)$ of random process β_n is to calculate its *characteristic function* $\varphi_B(t)$, which is defined by

$$\varphi_B(t) \equiv \overline{\exp[it\beta_n]} = \int_{-\infty}^{\infty} d\beta \, \rho_B(\beta) \exp(it\beta). \tag{C.1}$$

From the last equality in definition (C.1), it immediately follows that the desired probability density $\rho_B(\beta)$ is just the Fourier transform of $\varphi_B(t)$,

$$\rho_B(\beta) = \frac{1}{2\pi} \int_{-\infty}^{\infty} dt \, \varphi_B(t) \exp(-it\beta). \tag{C.2}$$

To start with, we substitute the explicit expression (3.68b) for β_n into definition (C.1) for characteristic function $\varphi_B(t)$. Then, we rewrite the result as an infinite product of exponential functions and take into account statistical independence of uncorrelated random variables α_n. This procedure yields

$$\overline{\exp[it\beta_n]} = \exp(it\overline{\beta}) \prod_{n'=-\infty}^{\infty} \overline{\exp[itG(n-n')\alpha_{n'}]}. \tag{C.3}$$

In accordance with the Gaussian distribution (3.69b) of α_n, its characteristic function is

$$\varphi_A(\tau) \equiv \overline{\exp(i\tau\alpha_n)} \equiv \int_{-\infty}^{\infty} d\alpha \, \rho_A(\alpha) \exp(i\tau\alpha) = \exp(-\tau^2/2). \tag{C.4}$$

The use of Eqs. (C.3), (C.4) with $\tau = tG(n-n')$, and normalization condition (3.4) results in

$$\varphi_B(t) = \exp(i\overline{\beta}t - t^2/2). \tag{C.5}$$

As is well known, this characteristic function corresponds to the Gaussian probability density (3.17). We can confirm this fact by direct evaluation of the integral in Eq. (C.2).

D. CORRELATOR OF γ_n

Let us derive the average $\overline{\gamma_n \gamma_{n+r}}$. Employing the standard integral presentation for the signum-function,

$$\operatorname{sign}(z) = \frac{1}{\pi} \int_{-\infty}^{\infty} dx \, \frac{\sin(zx)}{x}, \tag{D.1}$$

and Eq. (C.4) we arrive in a manner similar to calculation of characteristic function $\varphi_B(t)$ in Appendix 1, at the expression,

$$\overline{\gamma_n \gamma_{n+r}} = \mathcal{J}(K_\beta(r), \overline{\beta}) = \frac{2}{\pi^2} \int_0^{\infty} \frac{dx_1}{x_1} \int_0^{\infty} \frac{dx_2}{x_2} \exp\left(-\frac{x_1^2 + x_2^2}{2}\right)$$
$$\times \sum_{s=-1,1} s \exp[sK_\beta(r)x_1x_2] \cos[\overline{\beta}(x_1 - sx_2)]. \tag{D.2}$$

So, we have reduced the problem to deriving $\mathcal{J}(K, \beta)$. To solve it, we obtain the derivative of $\mathcal{J}(K, \beta)$ with respect to K and then make some simplifications,

$$\frac{\partial}{\partial K}\mathcal{J}(K,\beta)=\frac{1}{\pi^2}\int_0^{\infty}dx_1\int_{-\infty}^{\infty}dx_2\exp\left(-\frac{x_1^2+x_2^2}{2}\right)$$
$$\times\exp[Kx_1x_2]\sum_{t=-1,1}\exp[it\beta(x_1-x_2)].$$

To proceed, we write down the following known relation that is valid for arbitrary real quantities a and b,

$$\int_{-\infty}^{\infty}dx\exp(-x^2/2)\exp(ax)\exp(ibx)=\sqrt{2\pi}\exp[(a^2-b^2)/2]\exp(iab). \tag{D.3}$$

Using Eq. (D.3) with $a=Kx_1$ and $b=-t\beta$, we integrate over x_2 and then make some simplifications again,

$$\frac{\partial}{\partial K}\mathcal{J}(K,\beta)=\frac{\sqrt{2\pi}}{\pi^2}\int_{-\infty}^{\infty}dx_1\exp[-x_1^2(1-K^2)/2]$$
$$\times\exp(-\beta^2/2)\exp[i\beta(1-K)x_1].$$

Now we perform the change of integration variable x_1,

$$x_1'=x_1\sqrt{1-K^2}. \tag{D.4}$$

Then, applying Eq. (D.3) with

$$a=0,\qquad b=\beta\sqrt{\frac{1-K}{1+K}}, \tag{D.5}$$

we integrate over x_1 that gives rise to

$$\frac{\partial}{\partial K}\mathcal{J}(K,\beta)=\frac{2}{\pi\sqrt{1-K^2}}\exp\left(-\frac{\beta^2}{1+K}\right). \tag{D.6}$$

The general solution of Eq. (D.6) is

$$\mathcal{J}(K,\beta)=\mathcal{J}(0,\beta)+\frac{2}{\pi}\int_0^K\frac{dx}{\sqrt{1-x^2}}\exp\left(-\frac{\beta^2}{1+x}\right). \tag{D.7}$$

It should be noted that Eq. (D.7) can be also obtained by means of the two-point probability density that for correlated Gaussian sequence β_n with correlator $K_\beta(r)$ is defined by

$$\begin{aligned}\rho_B(\beta_n&=\beta,\,\beta_{n+r}=\beta')\\&=\frac{1}{2\pi\sqrt{1-K_\beta^2(r)}}\exp\left\{-\frac{\beta^2+\beta'^2-2K_\beta(r)\beta\beta'}{2[1-K_\beta^2(r)]}\right\}.\end{aligned} \tag{D.8}$$

The last step we should take, is to calculate $\mathcal{J}(0,\beta)$. It can be obtained directly from Eq. (D.2),

$$\mathcal{J}(0,\overline{\beta}) = \left[\frac{2}{\pi}\int_0^\infty dx \frac{\sin(\overline{\beta}x)}{x}\exp(-x^2/2)\right]^2 = \overline{\gamma}^2. \tag{D.9}$$

This result can be easily explained. Indeed, the condition $K_\beta(r)=0$ implies that correlations between β_n and β_{n+r} is lacking, hence, correlations between γ_n and γ_{n+r} is lacking as well.

Thus, as a result of all our calculations, we get

$$\mathcal{J}(K_\beta(r),\overline{\beta}) = \overline{\gamma}^2 + \frac{2}{\pi}\int_0^{K_\beta(r)} \frac{dx}{\sqrt{1-x^2}}\exp\left(-\frac{\overline{\beta}^2}{1+x}\right)$$

This expression provides Eq. (3.79).

E. UNIT-PULSE FUNCTION AND COMPLEMENTARY FUNCTION

Fist, it is useful to write down the canonical formulas for the Dirac delta-function and Kronecker delta-symbol,

$$2\pi\delta(k-k') = \sum_{r=-\infty}^{\infty}\exp[i(k-k')r] = 1+2\sum_{r=1}^{\infty}\cos[(k-k')r], \tag{E.1a}$$

$$2\pi\delta_{r,r'} = \int_{-\pi}^{\pi} dk\exp[i(r-r')k] = 2\int_0^{\pi} dk\cos[(r-r')k]. \tag{E.1b}$$

Then, in this Appendix we briefly consider two complementary functions that compose correlators and corresponding power spectra studying in Subsection 3.4.3,

$$D_{k_c}(r) = \frac{\sin(k_c r)}{k_c r}, \qquad r=0,\pm1,\pm2,\dots; \tag{E.2a}$$

$$\Delta_{k_c}(k) = \frac{\pi}{k_c}\Theta(k_c-|k|), \quad 0<k_c\leqslant\pi, \ |k|\leqslant\pi. \tag{E.2b}$$

The symbol k_c stands for the independent parameter, which specifies the variation scale k_c^{-1} of $D_{k_c}(r)$, and also, defines the value π/k_c and the position $k=\pm k_c$ of two symmetric jumps of $\Delta_{k_c}(k)$. By definition these functions obey the unit-normalization condition,

$$D_{k_c}(0) = \frac{1}{2\pi}\int_{-\pi}^{\pi} dk\Delta_{k_c}(k) = 1. \tag{E.3}$$

The introduced functions are the reciprocal Fourier transforms of each other,

$$D_{k_c}(r) = \frac{1}{2\pi}\int_{-\pi}^{\pi} dk\Delta_{k_c}(k)\exp(ikr); \tag{E.4a}$$

$$\Delta_{k_c}(k) = \sum_{r=-\infty}^{\infty} D_{k_c}(r)\exp(-ikr) = 1+2\sum_{r=1}^{\infty} D_{k_c}(r)\cos(kr). \tag{E.4b}$$

In the limiting case $k_c \to 0$, the first function tends to unity, while the second one turns into the Dirac delta-function,

$$\lim_{k_c \to 0} D_{k_c}(r) = \lim_{k_c \to 0} \frac{\sin(k_c r)}{k_c r} = 1\,; \tag{E.5a}$$

$$\lim_{k_c \to 0} \Delta_{k_c}(k) = \lim_{k_c \to 0} \frac{\pi}{k_c} \Theta(k_c - |k|) = 2\pi\delta(k). \tag{E.5b}$$

Otherwise, at $k_c = \pi$, the function $D_\pi(r)$ reduces to the Kronecker delta-symbol, whereas $\Delta_\pi(k)$ degenerates into one,

$$D_{k_c=\pi}(r) = \frac{\sin(\pi r)}{\pi r} = \delta_{r,0}\,; \tag{E.6a}$$

$$\Delta_{k_c=\pi}(k) = 1. \tag{E.6b}$$

We can note that, as a matter of fact, the function $\Delta_{k_c}(k)$ is a typical unit-pulse function and, as majority of them, our $\Delta_{k_c}(k)$ is also a prelimit Dirac delta-function. This fact is displayed by Eq. (E.5b).

References

1. *Dynamics and Thermodynamics of Systems with Long-Range Interactions,* T. Dauxois, S. Ruffo, E. Arimondo, and M. Wilkens, Lecture Notes in Physics Vol. 602 (Springer-Verlag, New York, 2002).
2. T. Padmanabhan, Phys. Rep. **188**, 285 (1990).
3. D. R. Nicholson, *Introduction to Plasma Theory* (John Wiley, New York, 1983).
4. J. Barre, T. Dauxois, G. De Ninno, D. Fanelli, and S. Ruffo, Phys. Rev. E **69**, 045501(R) (2004).
5. I. Kanter and D. A. Kessler, Phys. Rev. Lett. **74**, 4559 (1995).
6. G. K. Zipf, *Human Behavior and the Principle of least Effort* (Addison-Wesley, New York, 1949).
7. R. N. Mantegna, S. V. Buldyrev, A. L. Goldberger, S. Havlin, C.-K. Peng, M. Simons, and H. E. Stanley, Phys. Rev. E **52**, 2939 (1995).
8. Z. Ouyang, C. Wang, and Z.-S. She, Phys. Rev. Lett., **93**, 078103 (2004).
9. R. F. Voss, Phys. Rev. Lett. **68**, 3805 (1992).
10. H. E. Stanley *et. al.*, Physica A **224**, 302 (1996).
11. V. Pareto, *Le Cour d'Economie Politique* (Macmillan, London, 1896).
12. R. N. Mantegna, H. E. Stanley, Nature (London) **376**, 46 (1995).
13. D. Sornette, L. Knopoff, Y. Y. Kagan, and C. Vanneste, J. Geophys. Res. **101(B6)**, 13883 (1996); arXiv:cond-mat/9510035.
14. http://myhome.hanafos.com/philoint/phd-data/Zipf's-Law-2.htm.
15. C. Tsallis, J. Stat. Phys. **52**, 479 (1988).
16. *Nonextensive Statistical Mechanics and Its Applications*, eds. S. Abe and Yu. Okamoto (Springer, Berlin, 2001).
17. EHRENFEST P. Ehrenfest, T. Ehrenfest.-Encyklopädie der Mathematischen Wissenschaften.-1911.-Bd. IV 2, II.
18. P. Hänggi and P. Jung, Adv. Chem. Phys. **89**, 239 (1994).
19. S. O. Rice, Bell Syst. Tech. J. **23**, 282 (1944); S. O. Rice, in *Selected Papers on Noise and Stochastic Processes*, ed. by N. Wax (Dover, New York, 1954) p. 180.
20. D. Saupe, in *The Science of Fractal Images*, ed. by H.-O. Peitgen and D. Saupe (Springer, New York, 1988); J. Feder, *Fractals* (Plenum Press, New York, 1988).
21. A. Czirok, R. N. Mantegna, S. Havlin, H. E. Stanley, Phys. Rev. E **52**, 446 (1995).
22. A. Romero and J. Sancho, Journal of Computational Physics **156**, 1 (1999); J. García-Ojalvo and J. Sancho *Noise in Spatially Extended Systems*, (Springer-Verlag, New York, 1999).
23. F. M. Izrailev, N. M. Makarov, J. Phys. A: Math. Gen. **38**, 10613 (2005).
24. C. S. West and K. A. O'Donnell, J. Opt. Soc. Am. A **12**, 390 (1995).
25. F. M. Izrailev, A. A. Krokhin, Phys. Rev. Lett. **82**, 4062 (1999); A. A. Krokhin, F. M. Izrailev, Ann. Phys. (Leipzig) SI-**8**, 153 (1999).
26. R. Cakir, P. Grigolini, and A. A. Krokhin, Phys. Rev. E **74**, 021108 (2006).
27. B. B. Mandelbrot, Water Resour. Res. **7**, 543 (1971); J. Feder, Fractals (Plenum Press, 1988).
28. R. F. Voss, in *Fundamental Algorithms in Computer Graphics*, ed. R. A. Earnshaw (Springer, Berlin, 1985) p. 805.
29. W. Li, Europhys. Let. **10**, 395 (1989).

30. M. F. Shlesinger, G. M. Zaslavsky, J. Klafter, Nature (London) **363**, 31 (1993).
31. S. S. Melnyk, O. V. Usatenko, and V. A. Yampol'skii, Physica A **361**, 405 (2006).
32. O. V. Usatenko and V. A. Yampol'skii, Phys. Rev. Lett. **90**, 110601 (2003).
33. S. Hod and U. Keshet, Phys. Rev. E **70**, 015104(R) (2004).
34. S. L. Narasimhan, J. A. Nathan, and K. P. N. Murthy, Europhys. Lett. **69** (1), 22 (2005).
35. S. L. Narasimhan, J. A. Nathan, P. S. R. Krishna, and K. P. N. Murthy, arXiv:cond-mat/0409053.
36. See, for example, V. I. Klyatskin *Dynamics of Stochastic Systems* (Elsivier, 2005); P. A. Mello and N. Kumar *Quantum Transport in Mesoscopic Systems : Complexity and Statistical Fluctuations* (Oxford University Press, 2004).
37. N. M. Makarov, I. V. Yurkevich, *Zh. Eksp. Teor. Fiz.* **96** 1106 (1989) [*Sov. Phys. JETP* **69** 628 (1989)]; V. D. Freylikher, N. M. Makarov, I. V. Yurkevich, Phys. Rev. B **41**, 8033 (1990).
38. L. W. Rabiner, "A tutorial on hidden Markov models and selected applications in speech recognition," Proc. IEEE, 77(2), 257-285, February 1989.
39. F. Jelinek, Statistical Methods for Speech Recognition, MIT Press, Cambridge, MA, 1997.
40. R. Durbin, S. R. Eddy, A. Krogh, G. Mitchison, Cambridge University Press, 1999. ISBN 0-521-62971-3.
41. L. Pachter and B. Sturmfels, Cambridge University Press, 2005. ISBN 0-521-85700-7
42. K. Seymore, A. McCallum, and Roni Rosenfeld. Learning Hidden Markov Model Structure for Information Extraction. AAAI 99 Workshop on Machine Learning for Information Extraction, 1999.
43. J. Li, A. Najmi, R. M. Gray, Image classification by a two dimensional hidden Markov model, IEEE Transactions on Signal Processing, 48(2):517-33, 2000.
44. A. Krogh, I. S. Mian and D. Haussler, Nucleic Acids Research, 22(22), 4768–4778, 1994.
45. S. L. Salzberg, A. L. Delcher, S. Kasif and O. White, Nucleic Acids Research, 26(2), 544–548, 1998.
46. A. L. Delcher, D. Harmon, S. Kasif, O. White and S. L. Salzberg, Nucleic Acids Research, 27(23), 4636–4641, 1999.
47. A. Krogh, M. Brown, I. S. Mian, K. Sjölander and D. Haussler, J. Mol. Biol., **235**, 1501–1531, 1994.
48. S. E. Cawley, A. L. Wirth and T. P. Speed, Molecular and Biochemical Parasitology, **118**, 167–174, 2001.
49. W. H. Majoros and S. L. Salzberg, "An empirical analysis of training protocols for probabilistic gene finders," BMC Bioinformatics, available at http://www.biomedcentral.com/1471-2105/5/206, 2004.
50. L. R. Rabiner, A Tutorial on Hidden Markov Models and Selected Applications in Speech Recognition, Proceedings of the IEEE, **77(2)**, 257–286, 1989.
51. L. R. Rabiner, B. H. Juang, An introduction to hidden Markov models, IEEE ASSP Magazine, 4–15, 1986.
52. E. Charniak, Statistical Language Learning, IMIT Press. ISDN 978-0-262-53141-2, 1993.
53. Andrew J. Viterbi. IEEE Transactions on Information Theory, **13(2)**, 260–269, 1967.
54. G. D. Forney. The Viterbi algorithm. Proceedings of the IEEE **61(3)**, 268–278, 1973.
55. L. E. Baum, T. Petrie, G. Soules, N. Weiss, Ann. Math. Statist., V.**41**, N.1, 164–171, 1970.
56. J. Marbeau and S. Gudder, *Ann. Inst. Henri Poincare*, **52**, 31 (1990).
57. F. M. Izrailev, A. A. Krokhin, N. M. Makarov, O. V. Usatenko, Phys. Rev. E, **76**, 027701 (2007).
58. O. V. Usatenko, V. A. Yampol'skii, K. E. Kechedzhy and S. S. Mel'nyk, Phys. Rev. E **68**, 061107 (2003).
59. Z.A. Mayzelis, S. S. Apostolov, S. S. Mel'nyk, O. V. Usatenko, V. A. Yampol'skii, Chaos, Solitons and Fractals **34**, 112 (2007).
60. C. W. Gardiner: *Handbook of Stochastic Methods for Physics, Chemistry, and the Natural Sciences*, (Springer Series in Synergetics, Vol. 13 (Springer-Verlag, 1985).

61. A. Katok, B. Hasselblat: *Introduction to the Modern Theory of Dynamical Systems*, (Cambridge University Press; Factorial, Moscow, 1999) p. 768.
62. ftp://ftp.ncbi.nih.gov/genomes/bacteria /bacillus_subti-lis/NC_000964.gbk.
63. M. T. Madigan, J. M. Martinko, J. Parker: *Brock Biology of Microorganisms*, (Prentice Hall, 2002) p. 1104.
64. C. V. Nagaev, Theor. Probab. & Appl., **2**, 389 (1957) (In Russian).
65. I. A. Ibragimov, Yu. V. Linnik: *Independent and stationary connected variables*, (Nauka, Moscow, 1965) p. 524.
66. S. S. Apostolov, Z. A. Mayzelis, O. V. Usatenko, V. A. Yampol'skii, unpublished.
67. S. S. Apostolov, Z. A. Mayzelis, O. V. Usatenko, V. A. Yampol'skii, arXiv:physics/0610081, to be published in Int. J. Mod. Phys. B.
68. N. Metropolis, A.W. Rosenbluth, M.N. Rosenbluth, A.H. Teller, and E. Teller, J. Chem. Phys., **21**, 1087 (1953).
69. S. S. Melnyk, O. V. Usatenko, V. A. Yampol'skii, S. S. Apostolov, Z. A. Mayselis, J. Phys. A: Math. Gen. **39**, 14289 (2006).
70. U. Kuhl, F.M. Izrailev, A.A. Krokhin, and H.-J. Stockmann, Appl. Phys. Lett. **7**, 633 (2000).
71. J. A. Ogilvy, *Theory of wave scattering from random rough surfaces*, (Hilger, London, 1991).
72. E.L. Albuquerque, M.S. Vasconcelos, M. L. Lyra, and F.A.B.F. de Moura, Phys. Rev. E **71**, 021910 (2005).
73. P. Carpena, P. Bernaola-Galv̌an, P. Ch. Ivanov, H. E. Stanley, Nature **418**, 955 (2002); Nature **421**, 764 (2003).
74. R. A. Caetano and P. A. Schulz, Phys. Rev. Lett. **95**, 126601 (2005); A. Sedrakyan and F. Domínguez-Adame, Phys. Rev. Lett. **96**, 059703 (2006); E. Diaz, A. Sedrakyan, D. Sedrakyan, and F. Domínguez-Adame, Phys. Rev. B **75**, 014201 (2007).
75. See, e.g. N.G. Van Kampen *Stochastic Processes in Physics and Chemistry* (North Holland, Amsterdam, 1981).
76. T. Geisel, J. Nierwetberg, and A. Zacherl, Phys. Rev. Lett. **54**, 616 (1985).
77. P. Grigolini, Adv. Chem. Phys., **133**, 357 (2006).
78. S. S. Apostolov, A. A. Krokhin, N. M. Makarov, S. S. Melnyk, Z. A. Mayselis, O. V. Usatenko, to be published.
79. A. Erdelyi, Higher Transcendental Functions, V. 2, McGraw-Hill, 1953.
80. F. M. Izrailev, A. A. Krokhin, N. M. Makarov, S. S. Melnyk, O. V. Usatenko, and V. A. Yampol'skii, Physica A **372**, 279 (2006).
81. B. B. Mandelbrot, J. Van Ness, SIAM Review, **10**, No 4, 422 (1968)
82. F. Baldovin and E. Orlandini, Phys. Rev. Lett. **97**, 100601 (2006).
83. P. C. Hohenberg, B. I. Halperin, Rev. Mod. Phys., **49**, 435 (1977).
84. R. J. Glauber, J. Math. Phys. **4**, 294 (1963).
85. See, e.g., the special issue of Philosophical Transactions: Mathematical, Physical & Engineering Science **362**, No 1815 (2004); B. Ferguson and X.-C. Zhang, Nat. Mater. **1**, 26 (2002).
86. R. Kleiner, F. Steinmeyer, G. Kunkel, and P. Müller, Phys. Rev. Lett. **68**, 2394 (1992); G. Hechtfischer, R. Kleiner, A. V. Ustinov, and P. Müller, Phys. Rev. Lett. **79**, 1365 (1997); E. Goldobin, A. Wallraff, N. Thyssen, A.V. Ustinov, Phys. Rev. B **57**, 130 (1998); J. Zitzmann, A.V. Ustinov, M. Levitchev, S. Sakai, Phys. Rev. B **66**, 064527 (2002); M. Tachiki, M. Iizuka, K. Minami, S. Tejima, and H. Nakamura Phys. Rev. B **71**, 134515 (2005).
87. S. Savel'ev, V. Yampol'skii, A. Rakhmanov, F. Nori, Phys. Rev. B **72**, 144515 (2005); Physica C **437-438** 281 (2006); Physica C **445-448**, 175 (2006).
88. S. Savel'ev, A.L. Rakhmanov, F. Nori, Phys. Rev. Lett. **94**, 157004 (2005); Physica C **445-448**, 180 (2006); Phys. Rev. B **74**, 184512 (2006).
89. S. Savel'ev, V. Yampol'skii, and F. Nori, Phys. Rev. Lett. **95**, 187002 (2005); Physica C **445–448**, 183 (2006).
90. S. Savel'ev, A.L. Rakhmanov, V.A. Yampol'skii, and F. Nori, Nature Physics **2**, 521 (2006); J. Orenstein, Nature Physics **2**, 503 (2006).

91. See, e.g., H.S.J. van der Zant, T.P. Orlando, S. Watanabe, and S.H. Strogatz, Phys. Rev. Lett. **74**, 174 (1995); A. V. Ustinov, M. Cirillo, B.H. Larsen, V.A. Oboznov, P. Carelli, and G. Rotoli, Phys. Rev. B **51**, 3081 (1995); J. Pfeiffer, M. Schuster, A.A. Abdumalikov, Jr., and A.V. Ustinov, Phys. Rev. Lett. **96**, 034103 (2006); A.V. Ustinov, B.A. Malomed, and S. Sakai, Phys. Rev. B **57**, 11691 (1998).
92. J. Oppenlander, P. Caputo, Ch. Haussler, T. Trauble, J. Tomes, A. Friesch, and N. Schopohl, Appl. Phys. Lett. **83**, 969 (2003).
93. A.V. Ustinov and S. Sakai, Appl. Phys. Lett. **73**, 686 (1998).
94. H. Susanto, E. Goldobin, D. Koelle, R. Kleiner, S.A. van Gils, Phys. Rev. B **71** 174510 (2005).
95. J.D. Joannopoulos, R.D. Meade, and J.N. Winn, *Photonic Crystals* (Princeton University, Princeton, NJ, 1995).
96. N. Nishiguchi, S.I. Tamura, and F. Nori, Phys. Rev. B **48**, 14426 (1993); Phys. Rev. B **48**, 2515 (1993); S.I. Tamura, F. Nori, Phys. Rev. B **41**, 7941–7944 (1990).
97. V. Bellani, E. Diez, R. Hey, L. Toni, L. Tarricone, G.B. Parravicini, F. Domínguez-Adame,R. Gómez-Alcalá, Phys. Rev. Lett. **82**, 2159 (1999).
98. A.V. Ustinov, M. Cirillo, and B.A. Malomed, Phys. Rev. B **47**, 8357 (1993).
99. P. Anderson, Phys. Rev. **109**, 1492 (1958).
100. D.J. Thouless, J. Phys C **5**, 77 (1972).
101. D.J. Thouless, Phys. Rep. **13**, 93 (1974).
102. M. Griniasty and S. Fishman, Phys. Rev. Lett. **60**, 1334 (1988).
103. S. DasSarma, S. He, and X.C. Xie, Phys. Rev. Lett. **61**, 2144 (1988).
104. J.M. Luck, Phys. Rev. B **39**, 5834 (1989).
105. M. Kolár, M.K. Ali, F. Nori, Phys. Rev. B **43**, 1034 (1991).
106. D.H. Dunlap, H.-L. Wu, and P.W. Phillips, Phys. Rev. Lett. **65**, 88 (1990); P. Phillips and H.-L. Wu, Science **252**, 1805 (1991).
107. S. S. Melnyk, O. V. Usatenko, V. A. Yampol'skii, and V. A. Golick, Phys. Rev. E **72**, 026140 (2005).
108. www.dml.riken.go.jp/THz/localization.
109. S. Sakai, P. Bodin, N.F. Pedersen, J. Appl. Phys. **73**, 2411 (1993); M. Tachiki, M. Machida, Physica C **341-348**, 1493 (2000); S.N. Artemenko and S.V. Remizov, Physica C **362**, 200 (2001); Ch. Helm, L.N. Bulaevskii, Phys. Rev. B **66**, 094514 (2002).
110. V. Freilikher, E. Kanzieper, and A. A. Maradudin, Phys. Rep. **288**, 127 (1997).
111. F. M. Izrailev, N. M. Makarov, M. Rendón, Phys. Rev. B **73**, 155421 (2006).
112. D. B. Hunter and R. A. Minasian, Electr. Lett. **35**, 412 (1999).
113. T. Brandes, S. Kettemann (eds.), *Anderson Localization and Its Ramifications: Disorder, Phase Coherence, and Electron Correlations (Lecture Notes in Physics)* (Springer, Berlin, 2003); N.E. Hurt, *Mathematical Physics of Quantum Wires and Devices: From Spectral Resonances to Anderson Localization* (Springer, Berlin, 2000).
114. Q. Niu and F. Nori, Phys. Rev. Lett. **57**, 2057 (1986); Phys. Rev. B **39**, 2134 (1989); Phys. Rev. B **42**, 10329 (1990); F. Nori and J. P. Rodriguez, Phys. Rev. B **34**, 2207 (1986).
115. M. Kolár and F. Nori, Phys. Rev. B **42**, 1062 (1990).
116. V. Misko, S. Savel'ev, and F. Nori, Phys. Rev. Lett. **95**, 177007 (2005); V.R. Misko, S. Savel'ev, and F. Nori, Phys. Rev. B **74**, 024522 (2006); Physica C **437–438**, 213 (2006).
117. N.L. Frisch and S.P. Lloyd, Phys. Rev. **120**, 1175 (1960).
118. B. Halperin, Phys. Rev. **139**, A104 (1965).
119. J. C. Flores, J. Phys. Condens. Matter **1**, 8471 (1989).
120. A. Bovier, J. Phys. A **25**, 1021 (1992).
121. H. A. Makse et al., Phys. Rev. E **53**, 5445 (1996).
122. O. V. Usatenko, S. S. Melnyk, V. A. Yampolski, M. Johansson, L. Kroon, R. Riklund, Telecommunications and Radio Engineering, **66**, 353 (2007); arXiv:cond-mat/0507332.
123. V. I. Arnold and A. Avez, *Ergodic Problems in Classical Mechanics* (Benjamin, New York, 1968).

124. M. Reed and B. Simon, *Methods of Modern Mathematical Physics* (Academic Press, New York, London, 1978).
125. A. Sütϕ, Commun. Math. Phys. **111**, 409 (1987); J. Stat. Phys. **56**, 525 (1989).
126. G. Caldarelli, M. Marsili and Y.-C. Zhang, Europhys. Lett., **40** (5), 479 (1997).
127. B.A. Trubnikov, Priroda, **11**, 3 (1993) (in Russian).
128. B.B. Mandelbrot, in *Structure of Language and its Mathematical Aspects*, Proceedings of the Symposium on Applied Math, edited by Roman Jacobson (AMS, New York, 1961), vol. 12.
129. B.B. Mandelbrot, *The Fractal Geometry of Nature* (Freeman, New York, 1983).
130. W. Li, IEEE Trans. Inf. Theor. **38**, 1842 (1992); Complexity **1**, 6 (1996).
131. S. Denisov, Phys. Lett. A **235**, 447 (1997).
132. A. Schenkel, J. Zhang, and Y. C. Zhang, Fractals **1**, 47 (1993).
133. P. Kokol, V. Podgorelec, Complexity International **7**, 1 (2000).
134. W. Ebeling, A. Neiman, T. Poschel, arXiv:cond-mat/0204076.
135. The Old Testament of the King James Version of the Bible, http://www.writersbbs.com/bible/.
136. Russian Synodal LiO 31/7/91, http: //www.lib.ru/ hristian/bibliya/nowyj_zawet.txt.
137. http://www.lib.ru.
138. D. Lind and B. Marcus, *Symbolic Dynamics and Coding*, Cambridge Univ. Press, Cambridge, U.K., 1995.
139. http://www.southalabama.edu/mathstat/personal_pages/williams/.
140. G. D. Birkhoff, Memoriae Pont. Acad. Sci. Novi Lyncaei, s. 3, **1**, 85, (1935).
141. S. Katok and I. Ugarcovici, Bul.(New Series) Am. Math. Sci., **44**, 87 (2007).
142. J. Hadamard, J. Maths. Pures et Appliqu.es **4**, 27 (1898).
143. M. Morse and G. A. Hedlund, Amer. J. Math. **60**, 815 (1938).
144. C.E. Shannon, A mathematical theory of communication, Bell Syst. Tech. J. **27**, 379; 623 (1948).
145. I. P. Cornfeld, S. V. Fomin, and Ya. G. Sinai, *Ergodic Theory* (Springer, New York, 1982).
146. P. Góra, A. Boyarsky, *Laws of Chaos, Invariant Measures and Dynamical Systems in One Dimension*, Birkhäuser, Boston, 1997.
147. Annales de l'I.H.P., Sec. A, **62**, 251 (1995).
148. R. Adler and B. Weiss, Memoirs Amer. Math. Soc., **98** (1970).
149. Y.-Ch. Lai, E. Bollt, C. Grebogi, Phys. Lett. A **255**, 75 (1999).
150. A. Manning, TRANSACTIONS OF THE AMERICAN MATHEMATICAL SOCIETY, **354**, 2849 (2002).
151. C. L. Barrett and C. M. Reidys, Appl. Math. Comput. **98**, 241 (1999); C. L. Barrett, H. S. Mortveit, and C. M. Reidys, ibidem **107** 121 (2000); C. L. Barrett, H. S. Mortveit, and C. M. Reidys, ibidem **122**, 325 (2001).
152. Accardi, L. : Noncommutative Markov chains, School of Mathematical Physics, Camerino, 268–295 (1974)
153. http://en.wikipedia.org/wiki/Quantum_Markov_chain.
154. K.-H. Fichtner, L. Fichtner, Math. Inf. 08.05 (2005).
155. B.V. Gnedenko: *The Theory of Probability*, (Chelsea Publ. Co., 1962).

Index